The publisher and the University of California Press
Foundation gratefully acknowledge the generous support
of the Ralph and Shirley Shapiro Endowment Fund in
Environmental Studies.

America's Largest Classroom

America's Largest Classroom

What We Learn from Our
National Parks

EDITED BY

Jessica L. Thompson and Ana K. Houseal

with Abigail M. Cook

Foreword by Milton Chen

UNIVERSITY OF CALIFORNIA PRESS

University of California Press
Oakland, California

© 2020 by Jessica L. Thompson and Ana K. Houseal

Cataloging-in-Publication Data is on file at the Library of
Congress.

ISBN 978-0-520-34063-3 (cloth : alk. paper)
ISBN 978-0-520-34064-0 (pbk. : alk. paper)
ISBN 978-0-520-97455-5 (ebook)

Manufactured in the United States of America

29 28 27 26 25 24 23 22 21 20
10 9 8 7 6 5 4 3 2 1

For the next century of park visitors

Contents

List of Contributors xi

Foreword. National Parks: "America's Best"
  Outdoor Classrooms xiii
    Milton Chen

Preface xix

Acknowledgments xxi

SECTION I. THE LONG VIEW OF LEARNING IN THE PARKS 1

1 Dynamic Learning Landscapes: The Evolution of
   Education in Our National Parks 5
   Julia Washburn

2 Commentary: Perspectives on Heritage Leadership 23
   Theresa Coble

3 Invoking the Spirit of History on the Journey through
   Hallowed Ground 29
   James A. Percoco

4 Two Different Ways of Knowing the Glacier Area 37
   Donal Carbaugh

SECTION II. FEEDBACK LOOPS: SYSTEMS AND SCIENCE LEARNING 49

5 Learning about Climate Change in Our National Parks 53
Shawn Davis and Jessica L. Thompson

6 Place-Based Education at Teton Science Schools: Inspiring Curiosity, Engagement, and Leadership in National Parks and Beyond 73
Kevin Krasnow, Nate McClennen, Amanda Kern, Patrick Leary, and Greg Peck

7 Three-Dimensional Learning: "Upping the Game" in Citizen Science Projects 83
Ana K. Houseal

8 Mentoring Mountain Raingers: Beyond Basic Hydrological Field Research in the Great Smoky Mountains 97
Douglas K. Miller

SECTION III. HEALTH AND SELF: EMPOWERING LEARNING IN PARKS 111

9 Learning Environmental Psychology in the National Parks 115
Donna K. McMillan

10 Can Signage Influence Healthy Behavior? The Case of Catoctin Mountain National Park 127
Mallika Bose, Lara Nagle, Jacob Benfield, Heather Costigan, Jeremy Wimpey, and B. Derrick Taff

11 Learning Historic Places with Diverse Populations: An Exploratory Study of Student Perceptions 137
Jenice L. View and Andrea Guiden

12 "I Felt Like a Scientist!": Accessing America's National Parks on Every Campus 151
Natalie Bursztyn, Richard Goode, and Colleen McDonough

SECTION IV. PARTNERING FOR THE NEXT GENERATION OF LEARNERS 167

13 Place-Based Learning Fosters Engagement and Opportunities for Innovative Partnerships 171
Susan Newton

14   A Partnership Model of Education at Cuyahoga
     Valley National Park   183
     Deb Yandala, Katie Wright, and Jesús Sánchez

15   Pura Vida Inspires Diversity and Engagement at Grand
     Teton National Park   189
     Teddi (Hofmann) Freedman

16   What Really "Matters" at Stephen T. Mather Building
     Arts and Craftsmanship High School   195
     Deborah Shanley and Lois Adams-Rodgers

17   Learning Historic Places with Diverse Populations:
     Making the Case for Teacher-Ranger Professional
     Development   201
     Jenice L. View and Paula Cristina Azevedo

SECTION V. STRATEGIC INTENTION FOR PARK
LEARNING AND PRACTICE   217

18   Lessons Learned from Museums: Family Learning in
     National Parks   221
     Colleen Bourque and Ana K. Houseal

19   Identifying Outcomes for Environmental Education at
     National Parks   245
     Robert B. Powell, Marc J. Stern, and B. Troy Frensley

20   Valuing Education and Learning in the National Parks   259
     Tim Marlowe, Linda J. Bilmes, and John Loomis

21   Commentary: National Parks as Places for Free-Choice
     Learning   265
     Martin Storksdieck and John Falk

Afterword   271
     Jonathan B. Jarvis
Index   275

Contributors

Lois Adams-Rodgers, Independent Educational Leadership Consultant

Paula Cristina Azevedo, George Mason University

Jacob Benfield, The Pennsylvania State University

Linda Bilmes, Harvard University

Mallika Bose, The Pennsylvania State University

Colleen Bourque, University of Wyoming

Natalie Bursztyn, Quest University

Donal Carbaugh, University of Massachusetts Amherst

Milton Chen, Edutopia

Theresa Coble, University of Missouri-St. Louis

Heather Costigan, The Pennsylvania State University

Shawn Davis, Slippery Rock University

John Falk, Institute for Learning Innovation

Teddi (Hofmann) Freedman, University of Wyoming

B. Troy Frensley, University of North Carolina Wilmington

Richard Goode, Porterville College

Andrea Guiden, George Mason University

Ana K. Houseal, University of Wyoming

Jonathan B. Jarvis, Institute for Parks, People & Biodiversity, University of California, Berkeley

Amanda Kern, Teton Science Schools

Kevin Krasnow, Teton Science Schools

Patrick Leary, Wildlife Expeditions
John Loomis, Colorado State University
Tim Marlowe, Oakland Promise
Nate McClennen, Teton Science Schools
Colleen McDonough, California State University, Fullerton
Donna K. McMillan, St. Olaf College
Douglas K. Miller, University of North Carolina Asheville
Lara Nagle, The Pennsylvania State University
Susan Newton, University of Missouri-St. Louis
Greg Peck, Teton Science Schools
James A. Percoco, Loudoun School for Advanced Studies
Robert B. Powell, Clemson University
Jesús Sánchez, Cuyahoga Valley Environmental Education Center
Deborah Shanley, Brooklyn College of the City University of New York
Marc J. Stern, Virginia Tech
Martin Storksdieck, Oregon State University
B. Derrick Taff, The Pennsylvania University
Jessica L. Thompson, Northern Michigan University
Jenice L. View, George Mason University
Julia Washburn, Rock Creek Park
Jeremy Wimpey, Applied Trails Research
Katie Wright, Cuyahoga Valley Environmental Education Center
Deb Yandala, Conservancy for the Cuyahoga Valley National Park

National Parks

"America's Best" Outdoor Classrooms

MILTON CHEN

In the second decade of the twenty-first century, our American democracy lies at a crossroads. We face challenges to our economy, national security, and social fabric after decades of disruption from forces ranging from technology to terrorism. It's becoming clear that meeting these challenges depends on how well we educate more diverse generations of students for higher achievement. The research is definitive: college graduates earn more money, pay more taxes, avoid public welfare and incarceration, live healthier lives, and are more open- and civic-minded than their less-educated peers.

It's also becoming clear that the institution of traditional schooling, largely unchanged from previous generations, needs a bigger vision and innovative partners. The model of one teacher tasked with educating twenty-five students—within the four walls of the classroom, for 180 days a year—was created for an agrarian economy, when students needed to work on their family farms during the long summer vacation. The politics and policies of school districts often obstruct creative approaches to reforming teaching and learning.

It's high time we recognize that a missing key to unlocking the potential of our youth lies beyond the walls of the school building. A major part of the long-standing "achievement gap" can be explained by an "experience gap." Many of today's students, from all backgrounds, are growing up without the broad range of experiences that connect classroom lessons to real life and propel their educations forward with purpose and passion. The same digital technologies that can bring virtual field trips and global friendships to today's students are frequently overused for binge-watching TV and mundane texting.

In many coastal cities, it is astonishing to find so many high school students who live within five miles of the ocean but have never seen it. Many have never put their hands in dirt and grown a flower or a tree. Their generation faces dramatic environmental challenges, requiring solutions for clean air, water, and energy, but their "experience portfolio" needs rebalancing. As one Oakland teacher described her middle school boys on a field trip to a Bay Area national park, "These boys carry guns, but they're afraid of bugs."

Filmmaker Ken Burns, using novelist Wallace Stegner's line, titled his documentary on the national parks *America's Best Idea*. This extraordinary American idea—to set aside our nation's most inspiring landscapes and important historic sites, not for a wealthy aristocracy but for all members of society—can be a "best idea" for education as well. Our national parks are turning out to be our best outdoor classrooms, where students can literally "come to their senses" using all of their senses— their heads, hands, and hearts—to understand their place in the natural and human world.

Recent research, including findings from neuroscience, supports the engagement of students' bodies, as well as their minds, for deeper learning. In his 2013 book *Education and the Environment,* author and educator Gerald A. Lieberman reviewed research on the impact of place-based experiential learning on improved academic achievement, classroom behavior, preparation for college and careers, and personal confidence. Such programs also energized teachers and strengthened relationships between schools and communities.

As the National Park Service (NPS) enters its second century, it has joined with schools, universities, museums, libraries, nonprofits, and many youth-serving organizations to map a new ecosystem for learning. In more than four hundred national parks, from the iconic Yellowstone, Grand Canyon, and Yosemite, to the Nez Perce sites spanning four states, to the smaller jewels of the homes of César Chávez and Frederick Douglass, students are documenting flora, fauna, and human artifacts; restoring valuable habitats; and gaining a deeper understanding of our nation's conflicts. Through these place-based learning experiences, abstract ideas from biodiversity to cultural diversity come alive.

When students stand on the Gettysburg battlefield or at the USS Arizona memorial at Pearl Harbor, the phrase "hallowed ground" takes on new meaning. The NPS has committed to telling all of America's stories in the interpretation of current sites and development of new ones. Students gain a more complete picture of the impact of war by connecting Pearl Harbor to the unjust internment of Japanese-Americans at two other national parks, Manzanar and Tule Lake. Indeed, the themes of war, heroism, and injustice are fully evident in the Journey through Hallowed Ground program, described here by Jim Percoco, encompassing Gettysburg to Monticello, 180 miles, and thirteen national park units.

This book grew out of a 2016 meeting in Yosemite of the Education Committee of the National Park System Advisory Board. Our twenty-member group had been meeting for seven years, enthusiastic about the new directions the NPS was taking to elevate education as part of its core mission. This work was led by NPS Director Jon Jarvis, who appointed the first Associate Director for Interpretation, Education, and Volunteers, Julia Washburn.

A frequent theme of our meetings was the many inspiring ways in which learners of all ages were engaged in park-based experiential learning, but how little known these projects were. We highlighted some of these programs at the first-ever National Parks Learning Summit at National Geographic Society in Washington, DC, during the NPS Centennial in 2016. In reviewing that summit at the Yosemite meeting, we believed a book containing a larger collection of these programs and their related research would be invaluable. Two members of the committee, Dr. Jes Thompson and Dr. Ana Houseal, intrepidly agreed to take on this major project.

America's Largest Classroom represents, to my knowledge, the first compilation of stories and studies devoted exclusively to education in the national parks. These chapters embrace an astonishing spectrum of educational experiences for students, from K–12 through college. These studies also address adults, especially teachers and parents, and how national park experiences enhance their own knowledge and motivation and how they, in turn, can reinforce these outcomes for children.

Far beyond the typical field trip, this book documents how national parks are serving as partners with universities and nonprofits to create authentic, deeper, and more lasting learning. In immersive residential programs, such as the Cuyahoga Valley Environmental Education Center near Cleveland, students investigate an ecosystem of forests, meadows, and ponds and observe how shipping along the Ohio and Erie Canal contributed to the state's prosperity in the nineteenth century. NatureBridge offers such programs at six other national parks, including Yosemite, where students can walk in the footsteps of John Muir and the Buffalo Soldiers and learn how one founded the national park idea and the others protected it.

Park learning highlights collaborative learning, where students work in teams to accomplish larger projects, such as measuring water quality in New York Harbor or supporting their peers while biking eighteen hundred miles of the Underground Railroad. The Climate Change Academies operating at Indiana Dunes National Lakeshore and Cape Cod National Seashore allow high school students to spend two days learning about climate change monitoring through peat marshes, sea level rise, and bird and plant species, and then teach fourth graders on day three.

Teachers are often learning alongside their students, in partnership with interpretive park rangers. One of the best professional development programs, Teacher-Ranger-Teacher, involves teachers spending a summer in a national park and bringing home lesson plans to share with their students.

Many college faculty use the parks as outdoor laboratories. In this book, Dr. Douglas Miller of University of North Carolina describes his undergraduate field research using a rain gauge network in the Great Smoky Mountains, providing valuable data for NOAA and NASA. During the 2016 centennial, the NPS and the National Geographic Society expanded the popular BioBlitz, a citizen science project to conduct species inventories, to more than two hundred parks, cities, and campuses. These experiences enable students to work with others across age and racial boundaries and strengthen their social and emotional muscles, valuable assets as future team members in the workplace.

The national parks and their partners are employing an expanding array of educational media and technologies to engage students in novel ways, including a project described here where California State University students use augmented reality apps to learn the geology of the Grand Canyon. The NPS website (www.nps.gov/teachers) is full of lesson plans to whet students' appetites for deeper learning in the sciences, the environment, and their peoples.

Online field trips are connecting students to ranger talks from the Grand Canyon and underwater research at the Cabrillo National Monument. Webcams give students a bird's eye view of Alaskan brown bears catching salmon at Alaska's Katmai National Park or a fish's eye view of a kelp forest at California's Channel Islands. Google and others are producing virtual reality field trips to give the most immersive experience, short of actually going there, to forty national parks, from Alcatraz to the cliff dwellings of Montezuma National Park.

As our nation continues to urbanize, new national park initiatives are attracting more diverse city dwellers. Already, 75 percent of students live within fifty miles of a national park site. The first national park in Chicago, the Pullman National Historic Park, designated in 2015, commemorates George Pullman's innovations in rail travel and the first African American union. That journey continued to the White House, where the great-granddaughter of a Pullman porter is former First Lady Michelle Obama.

When students are transported to these places, their hearts follow their minds. They learn more, not only about the subject matter of science or history, but also about themselves, their interests, and their abilities. They engage in learning "from the outside in." They understand how the larger ecosystem is affected by their everyday actions in consuming water and plastic, gas and electricity. Students who have faced adversity in their lives see the parallels with endangered animals and plants and, taking lessons from those species, reflect on their own resilience. In his chapter, Dr. Donal Carbaugh of University of Massachusetts describes the practice of "deep listening" used by the Blackfeet Indians, drawing inspiration from their homeland, what is now Glacier National Park.

Importantly for policy makers, one chapter, authored by Dr. Linda Bilmes at Harvard's Kennedy School of Government and her colleagues, provides a creative and rigorous analysis of the economic value of these educational experiences, concluding that the value far exceeds the federal investment.

The St. Olaf College students who travelled to Rocky Mountain National Park for Dr. Donna McMillan's Environmental Psychology course say it best. As one put it, "Before this class, I thought I had a pretty good grasp on why the environment is important to our well-being. I really had no idea! From every book to discussion to assignment to adventure in nature, I learned something new every day and had my own previous conceptions challenged. . . . I understand my own love for the natural world so much better now."

I hope you'll enjoy *America's Largest Classroom* as much as I have. You might have similar reactions to the two I've had: one, every student should be able to have these experiences, and two, I'm still a learner and I'd like to have more of these experiences myself! Much like the national parks themselves, this book expands our vistas of how, at this pivotal time, education should be transformed and leads us to envision a new landscape of learning. Perhaps most importantly, it reminds us that, in the end, Mother Nature is still our most marvelous teacher.

Preface

This project started on the back of a napkin at the Yosemite Valley Lodge in mid-November 2016. Immediately following the US presidential election, members of the National Park Service Advisory Board's Education Committee met to discuss strategy for keeping education in the parks relevant in a changing political climate. On December 10, 2016, Congress passed the National Park Service Centennial Bill, which made education a mandate for the National Park Service (which historically only mandated preservation and protection since the Organic Act of 1916). Over the past one hundred years the ways visitors engage with and learn at National Park Service (NPS) sites has become increasingly sophisticated and rigorously evaluated. This edited collection presents some of the most recent research, strategies, and case studies celebrating the evolution of education in *America's Largest Classroom*—the national parks.

Today, the *largest classroom* includes 419 sites covering more than eighty-five million acres in every state, the District of Columbia, American Samoa, Guam, Puerto Rico, and the Virgin Islands. Every year hundreds of millions of people visit America's national parks, and millions of them engage in one of the NPS's many educational programs. Place-based education programs are not only about biodiversity, geography, and the environment; many of the national park sites engage visitors in reflective lessons about America's dark history. Such programs provide an immersive, reflective learning experience, recounting the diverse stories and struggles for equality and freedom in this country. We believe these conversations are as important today as ever before.

This book is an initial effort, on behalf of the National Park Service's Advisory Board's Education Committee, to compile some of the latest research and evidence from leading practitioners and scholars in the fields of informal learning, place-based education, STEM, digital technology, and educational partnerships. The Education Committee represents a diverse range of expertise and has been working diligently to advise strategy and research about education in (and *with*) the national parks for the next generation of citizens. Special efforts have been made to engage diverse audiences and discuss difficult issues, from race to climate change. This compilation provides snapshots of this exciting work, as the agency, academics, and partners collaborate to innovate and inspire lifelong learners across the country.

We designed this collection for educators, learners, managers, and partners seeking a broad and deep overview of the landscape of place-based education in our national parks. This collection provides multiple disciplinary perspectives and methodological approaches to place-based learning scholarship and practice. The diverse and rich case studies illustrate and synthesize key insights about the practice of place-based learning, and we hope this collection inspires future practitioners and scholars in the field.

Acknowledgments

This project would not have been possible without the vision and dedication of the former members of the National Park Service Advisory Board Education Committee. We are especially grateful to dozens of NPS staff for their courage, creativity, and diplomacy. Two charismatic champions helped to get this collection published: Dr. Milton Chen, senior fellow and executive director, emeritus, at the George Lucas Educational Foundation; and Jonathan Jarvis, former director of the US National Park Service and executive director for the Institute for Parks, People and Biodiversity at the University of California, Berkeley. The process for finalizing the content required extensive editing expertise, which we could not have done without the tireless efforts of Clare Gunshenan and Zoë Nelson at the University of Wyoming. We also had an outstanding editorial team at the University of California Press, led by Stacy Eisenstark, Robin Manley, Francisco Reinking, and Linda Gorman. Our deepest gratitude goes to Abigail Cook for volunteering, lifting up and energizing this project on so many fronts.

Visitors at the highest point in Rocky Mountain National Park at sunset. Courtesy of the National Park Service.

The Long View of Learning in the Parks

For more than a hundred years, visitors have been learning in our national parks. Park-based learning has a long history, punctuated with passionate champions and educators working together to create meaningful learning experiences. In this section, chapter authors examine the big picture of park learning; from institutional knowledge to interpersonal perspective sharing, there are many ways to know our country's landscapes and civic legacies. These authors explore the potential for learning, knowing, and listening in our parks.

Washburn's chapter, "Dynamic Learning Landscapes: The Evolution of Education in Our National Parks," chronologically explores the park learning movement that began over a century ago and continues today. The first documented national park educational programs began in 1886 at Yellowstone National Park. US Army infantrymen, deployed to protect the park, began to answer visitor questions and give what have become known as "ranger talks" at Old Faithful and the Upper Geyser Basin. Washburn eloquently chronicles the National Park Service's (NPS's) institutional journey to find a place for education in their mission, programming, and budget. Today the NPS hosts more than ninety thousand education programs annually, serving over seven million students and more than 300 million visitors.

Coble's keynote, "Commentary: Perspectives on Heritage Leadership," recognizes that our history is replete with slavery, discrimination, dispossession, violence, and racial cleansing. This chapter explores our tendency to whitewash history or give in to historical amnesia and how that affects our understanding of "heritage" in the United States. The author suggests that if we expand our gaze from national to global injustice and consider how our supply chains are linked to modern slavery and ecocide, then the opportunity for grief and growth abounds. And from there, deep learning can take root and nurture compassionate citizenship.

Percoco, a nationally recognized high school history teacher, chronicles a 180-mile field trip from Gettysburg to Thomas Jefferson's mountaintop home, Monticello, in Charlottesville, Virginia. This chapter, "Invoking the Spirit of History on the Journey through Hallowed Ground," describes how students explored thirteen national park units and numerous scenic rivers, roads, farms, and small towns during a school-sponsored bus trip. After the trip, students created interactive media projects to demonstrate their learning. They shared the projects in their communities and online, creating learning experiences for others— near and far.

Carbaugh's chapter, "Two Different Ways of Knowing the Glacier Area," introduces the power of perspective in place-based learning. First Peoples of our continent lived for centuries in areas that are now national parks. They lived in ways and on lands that were—and are— known deeply to them. These places are entwined with their traditional ways, a body of ancestral wisdom. This traditional understanding bears its own guidelines for being in places, being in parks. Through conversations with members of the Blackfeet tribe, the author uses an ethnographic approach and cultural discourse analysis to examine what it means to understand and know national parks today.

Historic photo of a ranger talk at Rock Creek Park, Washington, DC. Courtesy of the National Park Service.

1

Dynamic Learning Landscapes

The Evolution of Education in Our National Parks

JULIA WASHBURN

INTRODUCTION

For more than 100 years we have been using parks as dynamic learning landscapes. Renowned historian Dr. Robin Winks called the national parks "the single greatest university in the world" (McDonnell, 2001). Now with over 400 parks in all fifty states and US territories, this "university" hosts more than ninety thousand education programs annually, serving over seven million students. More than 300 million visitors flock to national parks each year, not just for recreation and scenery, but for learning. In fact, research shows that approximately 95 percent of learning takes place outside the formal classroom (Falk & Dierking, 2010), much of that happening in places like museums and parks through direct experiences. The National Park Service (NPS) may just be the largest network of informal or *free choice* learning venues in the world. This chapter chronologically explores the learning movement that began over a century ago in our parks and continues today.

EDUCATION: A FOUNDING PURPOSE OF NATIONAL PARKS

The first documented national park educational programs began in 1886 at Yellowstone National Park. US Army infantrymen, deployed to protect the park, began to answer visitor questions and give what became known as *cone talks* at Old Faithful and the Upper Geyser Basin. The infantrymen were the first park ranger–interpreters, sparking visitors' curiosity and awe. An 1888 letter to the editor of the *Livingston Enterprise* described one such talk: "The Corporal delivers a lecture similar to a man who is shooting off a magic lantern" (Bawden, 2013, para. 11). Park educators have been shooting off the "magic lanterns" of learning ever since.

Founding NPS Director Stephen Mather was fully invested in national parks' educational value, stating in 1917 that "one of the chief functions of the national parks and monuments is to serve educational purposes" (Mather, 1917, p. 7). Mather established the National Parks Education Committee in 1918 to promote educational activities. The committee was led by Robert Sterling Yard, a publisher and promoter of the NPS who became its first Chief of the Educational Division. Its objectives were to

- educate the public in respect to the nature and quality of the national parks;
- further the view of the national parks as classrooms and museums of nature;
- use existing publicity and educational systems so as to produce a wide result;
- combine in one interest the sympathy and activity of schools, colleges, and citizen organizations in all parts of the country; and
- study the history and science of each national park and collect data for future use (Pitcaithley, 2002).

When no funding was budgeted for Yard's salary, Mather demonstrated his commitment to national park education and paid Yard out of his own pocket. Due to the NPS's lack of resources, Yard went on to cofound what is now the National Parks Conservation Association, partially to help promote national parks for learning.

NATIONAL PARKS AND THE NATURE STUDY MOVEMENT

The late 1800s and early 1900s were also the height of the international nature study movement, which espoused an interactive approach to learning through direct observation rather than from a textbook. Nature study supporters believed that direct contact with nature had the ability to nurture affection for the environment and "bring joy in an industrialized world" (Lorsbach & Jinks, 2013, p. 9). Luminaries such as John Muir and Enos Mills also stimulated interest in learning about the natural world. Mills started nature guiding in the Rocky Mountains and called the national parks "the school of nature." He wrote,

Why not each year send thousands of school children through the national parks? Mother Nature is the teacher of teachers, these parks are the greatest of schools and playgrounds. No other school is likely so to inspire children, so to give them vision and fire their imagination. (Mills & Schmeckebier, 1917, p. 366)

Two of Mills's students, Esther and Elizabeth Burnell, became the first government licensed nature guides in a national park. They worked for a hotel but were approved by the NPS in 1917 (Danton, 1988, p. 5). This idea of experiential learning in real and beautiful places was at the heart of the beginning of national park education. After Mather observed nature study activities at Lake Tahoe, he was inspired to establish the Free Yosemite Nature Guide Service in 1920 (Russell, 1960). Thus began the NPS profession of park naturalist. Ansel Hall from Yosemite became the NPS's first chief naturalist in 1923 (Pitcaithley, 2002).

THE EMERGENCE OF INTERPRETATION AS AN EDUCATIONAL DISCIPLINE

A few years later, a group of national park superintendents adopted a resolution reinforcing the importance of parks as places for learning, stating that

the mission of the National Parks [Service] is to provide not cheap amusement, but healthful recreation and to supplement the work of schools by opening the doors of nature's laboratory to awaken an interest in natural science as an adjunct to the commercial and industrial work of the world. (Pitcaithley, 2002)

By 1925, with a more stable budget, the Park Service reestablished its Division of Education with Hall in the lead. At the recommendation of a 1929 NPS Educational Advisory Board, the division grew into the Branch of Research and Education under the leadership of zoologist Dr. Harold C. Bryant. Bryant built the division around four key principles, notably using the word *interpretation* when referring to educational programming for the general public:

- simple, understandable interpretation of the major features of each park to the public by means of field trips, lectures, exhibits, and literature;
- emphasis on leading the visitor to study the real thing itself rather than depending on secondhand information;

- utilization of highly trained personnel with field experience, able to interpret to the public the laws of nature as exemplified in all the parks and able to develop concepts of the laws of life, useful to all; and

- a research program that would furnish a continuous supply of dependable facts suitable for use in connection with the educational program (Pitcaithley, 2002).

During this period, the beloved national park *campfire talk* was born, along with many other types of park-based programming, including auto caravans, camera caravans, water cruises, nature and historic trails, exhibits, lectures, museums, libraries, college and university field studies, and a Junior Nature School designed specifically for children (Bryant & Atwood Jr., 1932). Support for park-based learning continued to grow along with the national park system, and in the 1930s the idea of interpretation began to take hold to describe the work of park naturalists, historians, and archaeologists who conducted educational activities with the public.

INTERPRETATION AND EDUCATION GROW

Congress codified NPS's role in historic preservation and historical education with the 1935 Historic Sites Act, directing the Secretary of the Interior to develop "an educational program and service for the purpose of making available to the public facts and information pertaining to American historic and archeological sites, buildings, and properties of national significance" (p. 2). This support was internal as well, evidenced by a 1940 Ranger Conference at the Grand Canyon, which recommended working with schools and teachers and starting outreach programs.

Progress toward building robust education and interpretation in the parks was put on hold during World War II. However, after the war, NPS Director Conrad Wirth began Mission 66, a ten-year effort to build out park infrastructure in time for the Service's fiftieth anniversary in 1966. The idea was to attract people to parks by making car access affordable to middle-class Americans. Mission 66 built many visitor centers, exhibits, museums, and other visitor amenities in parks across the country, expanding park-based learning to a new generation—the baby boomers (Davis, 2003).

During this time, writer, teacher, and philosopher Freeman Tilden began writing about national parks and became interested in the field of park interpretation and education. His 1957 seminal work *Interpreting Our Heritage* (reprinted in 1977) is perhaps the most influential book ever written about the profession, which simply and elegantly outlines six timeless principles. Tilden stressed the importance of inspiration and provocation as chief aims of interpretation, rather than merely a transfer of knowledge. He writes that "the purpose of Interpretation is to stimulate . . . a desire to widen [one's] horizon of interests and knowledge, and to gain an understanding of the greater truths that lie behind any statements of fact" (1977, p. 33). Tilden pushed interpreters to be guides in the search for meanings inherent in park resources. In his book *The Fifth Essence*, Tilden describes the idea of the soul of things beyond their tangible form. His work helped move park-based learning beyond the study of facts to personal relevance and inspiration, to exploration and discovery of meanings (1968).

To meet the learning needs of ever-growing visitation and also in preparation for its fiftieth anniversary, NPS established the Stephen T. Mather Training Center for interpretive rangers on the campus of the former historically black Storer College in Harpers Ferry, West Virginia. Later, the Service would colocate the Harpers Ferry Center for Interpretive Media next door.

THE ENVIRONMENTAL EDUCATION MOVEMENT

Another postwar educational shift was the movement from cataloging or naming objects and facts to identifying and describing the relationships among things in ecological systems, placing an emphasis on conservation. The 1960s marked the emergence of the environmental movement, and NPS Director George B. Hartzog embraced the idea of environmental education as a means to develop public understanding of the forces that shape the environment and to produce individual awareness of personal responsibility for environmental quality (Sherwood, 2011). In 1967, NPS Assistant Director for Interpretation Bill Everhart postulated that interpreting park resources alone was not enough, writing that

> we have not effectively carried out an educational campaign to further the general cause of conservation. . . . Only through an environmental approach to interpretation can an organization like ours, which has both Yosemite and the Statue of Liberty, achieve its purpose of making the park visitor's experience fully significant. (Mackintosh, 1986, chapter 3, "Environmental Interpretation" section)

In 1968, the Service worked with consultant Mario Mensesini to develop the National Environmental Education Development (NEED) program and materials for schools and parks. NEED had five strands of study: (1) variety and similarities, (2) patterns, (3) interrelation and interdependence, (4) continuity and change, and (5) adaptation and evolution. These strands were meant to be embedded in education and interpretive programming. The NEED strands were also taught as part of the environmental living program, introducing overnight immersion environmental education experiences in more than eighty parks at designated environmental study areas (Mackintosh, 1986).

Environmental education took on even more profound importance after Neil Armstrong's historic walk on the moon in 1969. Photos from space gave the public a new perspective on Earth as a unified and fragile planet—the effects of pollution were clearly visible. New public awareness about the environment was also spurred by Senator Gaylord Nelson's idea for Earth Day, a nationwide teach-in on April 22, 1970. This bipartisan effort led to the passage of the clean water, clean air, and endangered species acts, and the establishment of the Environmental Protection Agency. The NPS became involved by forming an environmental education task force to "expedite the establishment of an environmental education program that is integral to operations at all levels of the National Park Service—a program which will also assist public and private organizations concerned with the promotion of a national environmental ethic" (Mackintosh, 1986, chapter 3, "Environmental Interpretation" section). In 1972, the NPS Office of Environmental Interpretation was established. While the NEED program supported the use of national-level curriculum materials rather than programming that focused on park-specific themes, it is credited with placing a new focus on involving visitors and students as *participants*, instead of regarding them as passive spectators.

LIVING HISTORY

The concepts of involvement and immersion were taken a step further through *living interpretation*, also known as *living history*. The notion of "going back in time" became popular and reached an all-time high during the nation's bicentennial in 1976. However, some programs were geared more toward entertainment, were not historically accurate, or trivialized the historical events they were supposed to commemorate. In

1980, the Service issued careful guidelines for living history programs, ensuring their accuracy and relevance to park resources and themes. Still, the idea of bringing history to life through costumed interpretation, demonstrations, and museum theater remain viable and effective methods (Mackintosh, 1986).

INTERPRETATION AS A MANAGEMENT TOOL

The 1980s marked a return to local park-specific-themed interpretation, as well as a movement to use educational programming to accomplish management objectives, such as promoting safety, preventing vandalism, and encouraging resource protection behaviors. In his twelve-point plan, Director William Penn Mott, Jr. directed the Service to "stimulate and increase our interpretive and visitor service activities for greater public impact" (NPS, 1985, p. 9). As part of the movement to bring parks to people, Mott placed special emphasis on "urban recreational areas as major education centers" and encouraged parks to place their stories into the context of the values of the entire national park system and to quality of life (NPS, 1985, p. 9). Under the direction of Chief of Interpretation Michael Watson, the Service embraced *The Interpretive Challenge* based on the idea that "a visit to a National Park breaks the routine of life: it gives life context, value, and meaning. Among these natural wonders and historic settings learning becomes fun" (Raithel, 1989, p. 1). The Interpretive Challenge placed emphasis on professional excellence, evaluation, education, program integration, and media. The NPS's Washington office began publishing a quarterly journal titled *Interpretation*, and in 1989, for the first time, Interpretive Operations was reflected in a separate section of the NPS Management Policies. During this time, leadership of interpretation and education was embedded in the Division of Park Operations.

PARKS AS CLASSROOMS

To celebrate the seventy-fifth anniversary of the NPS in 1991, Director James Ridenour convened a symposium in Vail, Colorado. The subsequent publication, *National Parks for the 21st Century: The Vail Agenda*, made sweeping recommendations for interpretation and education, asserting that "the Service should revise its philosophy, policy, and management approaches to reflect the legitimate role the agency has as a national public education system" (NPS, 1991, p. 89). The Vail Agenda also placed a specific emphasis on reaching a diverse public both in parks and in communities, using communication technologies and developing a "complete K–12 curriculum for school teachers to integrate the national parks into the classroom" (NPS, 1991, p. 90). This would manifest as the Parks as Classrooms program.

Parks as Classrooms was in part shaped by a 1993 task force report, released under the leadership of Chief of Interpretation Corky Mayo, which called for improving public education by assisting teachers with innovative educational methods (Chief, Division of Interpretation, 1993, p. 1). Diverging slightly from the Vail Agenda, Parks as Classrooms asserted that the best education programs were built through local partnerships between parks and schools. Further, it held that programs should be cocreated *with* teachers if they were going to be relevant and sustained. For the first time, a dedicated competitive funding source was established to support parks in developing curriculum-based education programs. With ever-tightening budgets and more constraints on schools, school administrators could not justify out-of-school field trips unless these activities aligned with curriculum objectives. Rather than a national-level parks curriculum, individual parks worked collaboratively with local or state school systems to ensure park programming aligned with these school curricula. Two influential field interpreters, Patti Reilly and Kathy Tevya, developed a national course to train park educators on developing and managing curriculum-based programs, which included methods for creating teacher professional development experiences. These practices are still in place today.

The 1993 report also called for fundraising and mass marketing to promote parks for learning. The congressionally chartered nonprofit partner of the NPS, the National Park Foundation (NPF), embraced Parks as Classrooms, fundraising millions of dollars to create grants and support educational programming in parks. One grant, Parks as Resources for Knowledge in Science (PARKS), a partnership with the National Science Teachers Association, sought to help integrate the National Science Education Standards into programs. The NPF also invested in evaluation, bringing a new level of rigor and documentation of participant outcomes to programming. For example, the PARKS program reached ninety-thousand students, who showed increases in stewardship for national parks and higher levels of perceived science learning. The majority of teachers participating indicated the program had given them new ideas to incorporate into their own science teaching (Wiltz, 2001). Program evaluation gives the NPS information necessary to improve programs over time and to better tailor future initiatives to desired participant outcomes.

THE INTERPRETIVE (R)EVOLUTION

In the early 1990s, the field of interpretation in the NPS underwent an identity crisis. Leaders felt the profession was unfocused and ill-defined, resulting in a renewed focus on interpretive training and an endeavor to evolve the profession. Under Mayo's leadership, along with David Dahlen and David Larsen of the Mather Training Center and hundreds of field interpreters, the Interpretive Development Program (IDP) was established. The IDP was a performance-based competency program (the first of its kind for the field of interpretation) that set professional standards, highlighting interpretation's ability to connect the interests of the visitor and the meanings of the resource. One basic tenant of the IDP was "the visitor is sovereign." In other words, in the tradition of Tilden, it was not the interpreter's job to tell people what to think or feel, rather to inspire the visitor to develop their own meanings and connections—both emotional and intellectual. While the focus of the IDP was still on the interpreter developing the ideas and delivering the programs, this new philosophy laid the groundwork for the current movement to audience-centered interpretation and human-centered programming (NPS, 1998).

NPS ONLINE

Parknet was established in 1994 as the NPS's first online web presence. It was soon embraced as the busiest "visitor center" in the NPS, opening up a whole new audience for park programs. Parknet would evolve into nps.gov, one of the most visited websites of the US government. Online NPS resources for teachers would also grow with an Education Portal. This site was the entry point for thousands of lesson plans, educational activities, videos, virtual park experiences, webcams, an online junior ranger program, and a plethora of social media sites. Through a partnership with the National Council for Social Studies, the NPS developed Teaching with Historic Places, which paired park rangers with teachers to create park-based history lessons for classroom instruction. Teaching with Historic Places is now a robust website of history education resources.

THE ROLE OF PARTNERS

A 1997 Education Initiative Symposium, convened in Santa Fe, brought leaders in the field of education, including park partners, together with NPS staff to consider the future of NPS Education—and collaboration with partners was key (NPS, 1998). Collaboration with partners has been central to interpretation and education, going back to the early days when Mather engaged the help of organizations such as the Smithsonian and the American Association of Museums to help establish national parks as places for learning. Today, a system of more than seventy nonprofit cooperating associations run park bookstores across the country and support interpretation and educational programming and research. Cooperating associations play an integral role in park interpretive operations. Numerous partners, such as Nature Bridge and the Cuyahoga Valley Conservancy, run park-based residential education centers and provide interpretive and educational programs in parks along with thousands of volunteers.

THE TURN OF THE CENTURY

With a new century on the horizon, Director Robert Stanton hosted a major national park conference, Discovery 2000, to discuss the future of parks. The conference highlighted interpretation and education as well as branding and marketing. Stanton also asked the National Park System Advisory Board to examine the NPS in the context of a changing society. The result was the 2001 report *Rethinking the National Parks for the 21st Century* (National Park System Advisory Board, 2001). The Board's Education Committee, led by renowned historian Dr. John Hope Franklin, called for the Service to "embrace its mission as educator" and take its place as part of the nation's education system (National Park System Advisory Board, 2001). As a result, the NPS National Leadership Council embraced a six-month seminar series on park-based learning, organized and strongly influenced by NPS Chief of Policy Loran Fraser. The result was a report titled *Renewing Our Education Mission* (NPS, 2003). The report is significant in that the entire national leadership of the NPS embraced an education vision together and connected park-based experiences to promoting the principles of democracy:

> Interpretation and education is a primary organizational purpose of the National Park Service, essential to achieving our mission of protecting and preserving our nation's natural and cultural resources. We envision a

national park system that is recognized as a significant resource for learning, where people and organizations collaborate on teaching and learning about the interconnections of human culture and nature, natural systems, the values of America's diverse heritage, and the principles of democracy. Parks are an integral part of the nation's educational system providing unique and powerful individual learning experiences that help shape understanding and inspire personal values. (NPS, 2003, p. 2)

THE INTERPRETATION AND EDUCATION RENAISSANCE

Among its recommendations, *Renewing Our Education Mission* called for a business plan for education and chartered a National Education Council, a cross section of NPS employees at all levels of the Service to advise the leadership body (NPS, 2003, pp. 7, 10). The National Education Council produced the first ever data-driven business assessment and plan for interpretation and education which documented funding and staffing trends by region (NPS, 2004). The business plan data, along with projected demographic changes, supported strategies laid out in the 2006 Interpretation and Education Renaissance Action Plan (NPS, 2006a, p. 6). The prescribed plan was based on five pillars:

- connecting *all* Americans to their national parks, including "ethnic, socioeconomic, and disabled groups that have . . . not been well connected to national parks" (NPS, 2006a, p. 6), with a focus on collaborating *with* audience groups, rather than creating programming *for* them (a nod toward inclusive, audience-centered programming);
- using new technologies—on the dawn of the smartphone the Service recognized that it must embrace technology to stay relevant rather than resist it;
- embracing education partners—there was a realization that as many as seventy thousand volunteers and partners were delivering interpretation and education services as compared with approximately four thousand uniformed park interpretive rangers and guides;
- developing and implementing professional standards, building on the progress of the IDP; and
- creating a culture of evaluation—increasingly focusing programs on outcomes and making social science data-driven decisions around staffing and programming.

One of the major outcomes of the Interpretation Renaissance was a Servicewide Evaluation Summit, sponsored by the National Park Foundation, the National Education Council, and the Advisory Board Education Committee. The summit's goal was to "generate useful dialogue about creating a culture of evaluation within Interpretation and Education characterized by continuous learning and decision making based on audience analysis and outcome data" (NPS, 2006b, p. 1). The entire NPS leadership, representatives from across the field and partner groups, and three former NPS directors attended the summit. Never before had such a large and comprehensive representation of the NPS leadership and field gathered to think collectively and exclusively about interpretation and education and how to measure it. A focus on asking the right questions, holding people accountable for learning, exploring the role of technology in place-based learning, prioritizing cultural competence, and evaluating visitor experiences were some of the themes explored. As a capstone, using new technology, the National Leadership Council held a live video conference with park staff across the country to share the ideas that emerged from the summit (NPS, 2006b).

THE NPS CENTENNIAL: EMBRACING A LARGER ROLE

Leading into the 2016 centennial of the NPS, the National Parks Conservation Association convened a Second Century Commission of prominent citizens to make recommendations to the president, Congress, and the NPS for a second century of parks. Their report, *Advancing the National Park Idea* (National Parks Conservation Association, 2009a), emphasized education and learning and connecting people to parks as key to the survival of national parks in a new century. The commission also emphasized the value of parks to the nation in building human capital and supporting democracy. Facing rapidly changing technologies, major demographic shifts, and the reality of climate change, the commission focused on the need for parks to be truly inclusive and relevant to all people's lives, promoting science literacy and civic engagement. Among its recommendations, the commission called for new legislation that "clearly secures its educational mission for the second century" (National Parks Conservation Association, 2009b). The commission also called for reestablishing a senior executive-level position to lead Interpretation and Education, elevating the department's role within the structure of the Service.

Director Jon Jarvis embraced the commission's recommendations and appointed many of the commissioners to the NPS Advisory Board, including Dr. Milton Chen of the George Lucas Educational Foundation to lead the education committee. Jarvis emphasized the immense value of the national park system to the nation for ecological preservation, health, economic prosperity, and education.

The Service's 2011 *A Call to Action* report, which called for a second century of stewardship and engagement, focused extensively on connecting people to parks, education, and community engagement. *A Call to Action* espoused an expansive vision:

> In our second century, we will fully represent our nation's ethnically and culturally diverse communities. To achieve the promise of democracy, we will create and deliver activities, programs, and services that honor, examine, and interpret America's complex heritage. By investing in the preservation, interpretation and restoration of the parks and by extending the benefits of conservation to communities, the National Park Service will inspire a "more perfect union," offering renewed hope to each generation of Americans. (NPS, 2011, p. 5)

A Call to Action challenged the Service to "embrace a larger education role, building an understanding of our country's shared heritage and preparing American citizens for the duties and responsibilities of citizenship" (NPS, 2011, p. 13). A theme of Jarvis's tenure was parks as *places for learning*. He appointed an Associate Director for Interpretation, Education, and Volunteers, and all those years after becoming a junior ranger, I was honored to take on that challenge.

Together with national park education partners, the National Park Foundation, a strong National Council for Interpretation Education and Volunteers, the Mather Training Center, regional chiefs of interpretation, the National Park System Advisory Board Education Committee, scholars from a variety of universities, field interpreters, and many more, we embraced the challenge of the centennial. We focused our efforts on relevance and inclusion, placing interpretation of events and ideas in their greater context, climate change education, and developing interpretive skills for the twenty-first century. With the help of author Lotte Lent from George Washington University, the Advisory Board Education Committee published an Interpretive Skills Vision Paper which laid out three targeted goals for interpretation (Lent, 2014, p. 3):

> **Audience Desires:** to facilitate meaningful, memorable experiences with diverse audiences so that they can create their own connections (onsite and virtually) with park resources

NPS Mission: to encourage shared stewardship through relevance, engagement, and broad collaboration

Societal Needs: to support global citizens to build a just society through engagement with natural and cultural heritage, by embracing the pursuit of lifelong learning

PART OF THE EDUCATIONAL ECOSYSTEM

With inspiration from Dr. John Falk of Oregon State University, the Advisory Board envisioned the NPS as part of a complex *educational ecosystem* consisting of formal and informal education facilitators—schools, libraries, museums, zoos, aquaria, universities, and the media—where learning happens 24/7/365, anywhere, life-wide and life-deep, with interpreters as facilitators of learning. This shift squarely places learners at the center of programming and establishes a collaboration between the learner and the interpreter. As a result, through the leadership of Katie Bliss at the Mather Training Center, a new set of interpretive competencies are emerging for a new century of learning. A new five-year strategic plan, *Achieving Relevance in Our Second Century*, seeks to put the vision paper's goals into action and reflects the ongoing evolution of interpretation through an emphasis on relevance and inclusion, educational leadership, active engagement, and business acumen (NPS, 2014).

A CENTENNIAL ACT

On December 16, 2016, President Obama signed the National Park Service Centennial Act into law. Title three section 301 states that

> the Secretary shall ensure that management of System units and related areas is enhanced by the availability and use of a broad program of the highest quality interpretation and education.

This law firmly establishes interpretation and education as central to the mission and purpose of the NPS and underscores its importance in the future of the National Park System and the nation. Why? As stated in *Achieving Relevance* and in the spirit of one hundred years of park-based learning, to

> leave the world in a better place . . . to increase quality of life, help realize the vision of democracy, ensure that all Americans' stories are shared, improve education and health, and support environmental and institutional adaptation and resilience. (NPS, 2014, p. iii)

With people at the center, true collaboration with partners and learners, and the interpreter as facilitator, we can realize this vision together.

References

Bawden, F. S. (2013, November 7). The immortal 15, Fort Yellowstone, and buffalo soldiers (part 1 of 2). Retrieved from https://www.nps.gov/yell/blogs/the-immortal-15-fort-yellowstone-and-buffalo-soldiers-part-1-of-2.htm

Bryant, H. C., & Atwood, W. W., Jr. (1932). *Research and education in the national parks*. Washington, DC: Government Printing Office.

Chief, Division of Interpretation. (1993, August 16). *Education strategy paper*. Washington, DC: National Park Service.

Danton, T. (1988). Before Tilden: A profession emerges. *Interpretation,* Spring, 3–7.

Davis, T. M. (2003). Research report: Mission 66 initiative. *CRM: Journal of Heritage Stewardship, 1*(1), 97–101.

Falk, J. H., & Dierking, L. D. (2010). The 95 percent solution. *American Scientist, 98*(6), 486–493.

Historic Sites Act, 49 Stat. 666 (1935).

Lent, L. (2014). Vision paper: 21st century National Park Service interpretive skills. Retrieved from http://goo.gl/WE7Exi

Lorsbach, A., & Jinks, J. (2013). What early 20th century nature study can teach us. *Journal of Natural History Education and Experience, 7,* 7–15.

Mackintosh, B. (1986). Interpretation in the National Park Service: A historical perspective. Retrieved from https://www.nps.gov/parkhistory/online_books/mackintosh2/index.htm

Mather, S. T. (1917). Annual report of the superintendent of national parks to the secretary of the interior for fiscal year ending June 30, 1916. Retrieved from http://npshistory.com/publications/annual_reports/director/1916.pdf

McDonnell, J. A. (2001). The National Park Service looks toward the 21st century: The 1988 general superintendent's conference and discovery 2000. Retrieved from https://www.nps.gov/parkhistory/hisnps/NPSThinking/suptreport.htm

Mills, E. A., & Schmeckebier, L. F. (1917). *Your national parks*. Boston, MA: Houghton Mifflin.

National Park Service. (1985). *12-point plan: The challenge.* Arizona Memorial Museum Association and the National Park Foundation. Retrieved from http://npshistory.com/publications/management/12-point-plan-challenge.pdf

———. (1991). *National parks for the 21st century: The Vail agenda.* Montpelier, VT: Capital City Press.

———. (1998, July). *Findings and recommendations education initiative symposium.* Washington, DC: Government Printing Office.

———. (2003). Renewing our education mission: Report of the national leadership council. Retrieved from https://www.nps.gov/archeology/aiassess/2003Educ_Mission.pdf

————. (2004). *Interpretation and education program—Business plan: Fiscal year 2004.* Washington, DC: National Park Service Education Council.

————. (2006a). Interpretation and education renaissance action plan. Retrieved from https://www.nps.gov/training/tel/Guides/Interp_Ed_Action_Plan_pg_20061117.pdf

————. (2006b). Learning together: Proceedings, evaluation, and applying lessons learned. Retrieved from https://www.nps.gov/archeology/aiassess/Learn_together_ExecSum2006.pdf

————. (2011, August 25). A call to action: Preparing for a second century of stewardship and engagement. Retrieved from https://www.nps.gov/calltoaction/PDF/Directors_Call_to_Action_Report.pdf

————. (2014). Achieving relevance in our second century. Retrieved from https://www.nps.gov/getinvolved/upload/IEVStrategicPlan_FINAL.pdf

National Park Service Centennial Act, H.R. 4680 (2016).

National Park System Advisory Board. (2001). Rethinking the national parks for the 21st century. Retrieved from https://www.nps.gov/policy/report.htm

National Parks Conservation Association. (2009a). Advancing the national park idea: National parks second century commission report. Retrieved from https://www.nps.gov/civic/resources/commission_report.pdf

————. (2009b). Education and learning committee report. Retrieved from https://www.npca.org/resources/1900-national-parks-second-century-commission-report

Pitcaithley, D.T. (2002). National parks and education: The first twenty years. Retrieved from https://www.nps.gov/parkhistory/resedu/education.htmRaithel, K. (1989). The interpretive challenge [Special issue]. *Interpretation*, Fall, 1–28.

Russell, C. (1960). A 40th anniversary. *Yosemite Nature Notes,* 39(7), 3–5.

Sherwood, F.B. (Ed.). (2011). *George B. Hartzog, Jr.: A great director of the National Park Service.* Clemson, SC: Clemson University Digital Press.

Tilden, F. (1968). *The fifth essence: An invitation to share in our eternal heritage.* Washington, DC: National Park Trust Fund Board.

————. (1977). *Interpreting our heritage.* Chapel Hill: University of North Carolina Press.

Wiltz, K.L. (2001) Parks as resources for knowledge in science: National program evaluation report. National Park Foundation. Retrieved from http://www.academia.edu/7597207/Parks_as_Resources_for_Knowledge_in_Science_evaluation, 15-19.

Lincoln Memorial. Courtesy of the National Park Service.

2

Commentary: Perspectives on Heritage Leadership

THERESA COBLE

John Rawls, a social ethicist, wondered what kind of society would you create if you were given the role of *Grand Designer*? There would be one catch, however. After you established your hierarchies and social stratifications, after you put in place pathways and barriers to social mobility, after you contextualized privilege and marginalization on the land, after you institutionalized justice to be more or less blind according to your preconceived notions of worth, you would be randomly assigned a niche in society, whether prominent or forlorn. Essentially, you would design the society and then would be assigned to an unknown gender, race, income, role, and position (Sandel, 2009). If those were the stakes, would you design a society like the one we have now? If you answer "no," then you are staring down the crux of heritage.

While we rightfully celebrate and affirm our ideals, aspirations, heroes, and accomplishments, we also tend to whitewash history or give in to historical amnesia. Our history is replete with slavery, discrimination, dispossession, violence, and racial cleansing. When the power of state, or the relative political power of individuals, was used to reinforce disparities, you or I might find occasion to grieve. If we expand our gaze from national to global injustice and consider how our supply chains are linked to modern slavery and ecocide (Bales, 2016), then the opportunity for grief and growth abounds. This is the context within which heritage leadership operates.

The etymology of the word *apologist* comes from the Greek word *apologia*, meaning "speaking in defense." Apologists often speak in defense of unpopular beliefs or ideas. Angry, arrogant, and bitter apologists retrench. They circle their wagons and create rival factions; they respond with shrill diatribes and have a dampening effect on public discourse.

I see a heritage apologist, however, as one who debunks myths. A heritage apologist listens critically to truth claims as they rebound within societal echo chambers. They are able to isolate information of dubious accuracy, hold it up for further scrutiny, and point the way to productive engagement. They reveal the meanings and significance of the past in all its complexity, ignominy, and glory. They make the past—and other unfamiliar realms—relevant to the people and issues of the day. We need heritage apologists—who call us collectively to higher ground—and we need a lot of them.

Heritage isn't tangible, but its effects are. Heritage is a lens through which we see and comprehend our world. It engages the sense-making part of our brain, helping us weave a narrative about our purpose, a narrative that clarifies if we belong and to whom. Heritage undergirds conclusions we draw about why our life matters, which memories to cherish, and which ideals to uphold. Heritage is a set of frequently unquestioned assumptions about why, individually and collectively, we have succeeded or failed to succeed.

Heritage is commonly defined as something handed down from the past. Lowenthal (1998) highlights that heritage is not history at all: "It is not an inquiry into the past, but a celebration of it . . . a profession of faith in a past tailored to present-day purposes" (p. x). Heritage encompasses the rituals, technology, social relations, and social movements that exert a powerful pull on our loyalties and sentiments. Heritage helps us make sense of the world around us—but it preconditions what we see and what we fail to see; it binds and blinds. We celebrate heritage, but always in part. We exhibit heritage, when and how it suits us. And much of heritage falls into the realm of the taken for granted—both binding and blinding in ways that elude conscious thought.

When we engage heritage, we face a perception crux. That is, we struggle to understand how the things handed down from the past influence the present. We struggle to see connections. Jaynes (1976) tells us that "language is an organ of perception, not simply a means of communication" (p. 50). Thus, overcoming this perception crux requires us to wrap words around our own experiences and the experiences of others, past and present. Each of us must articulate our conclusions, however tentative, about our place, our environment, our world—our niche—and our responsibility to the whole. If our words are consistent with what an honest inquiry into the past teaches us, we are well poised to act effectively in the present.

To negotiate the perception crux, we have to continually enlarge our capacity to see the whole. Heritage leaders facilitate this process for themselves and others with equal measures of humility and courage. Humility is appropriate because we will always have a limited ability to see the whole. In many ways, the hard wiring in our brains, our cognitive habits of mind, and our implicit biases threaten to derail the process (Banaji & Greenwald, 2013; Fine, 2008; Marshall, 2015). Courage is required because the whole is not, and has never been, pretty. But seeing the whole, and understanding how all the parts connect, sets us up to become a whole person. This mirrors the challenge Leopold (1949) identified for those engaged in the field of outdoor recreation: "Recreational development is a job not of building roads into the lovely country, but of building receptivity into the still unlovely human mind" (pp. 176–7).

Before we can foster receptivity to the tough truths of history, before we can engage audiences about how to enhance equity in the present, we have to create safe spaces to converse about our perceptions, understandings, and lived experiences. Patterson, Grenny, McMillan, and Switzler (2011) tell us that the pressure cooker dial notches up whenever the stakes are high, emotions are engaged, and opinions vary. Holding conversations in these conditions can feel like navigating a minefield. Worse, heritage will eventually bring every one of us to discrepancy. That is, the way we think the world should work, our global meanings, and the meanings we ascribe to a specific situation will not line up. Park (2010) tells us that discrepant life events cause distress, pushing us toward the hard work of meaning making. When our global and situational meanings realign, we may experience many of the following outcomes: restored/changed sense of meaning in life, changed global goals and beliefs, reappraised meaning of the stressor, changed identity, perceptions of growth or positive life changes, reattributions or causal understanding, acceptance, and a sense of having "made sense."

Heritage leaders help people "make sense" in difficult and varied situations. They help people experience discomfort, empathy, or other strong emotions, because doing so can facilitate the formation of new meanings. They help people explore how these new meanings change the way they view themselves and the world around them. In short, heritage leaders help people heal, grow, and transform.

Resolving the tensions embedded in heritage requires us to draw ever-wider circles, with each successive arc circumscribed by love. Heritage leaders engage all who will leave hate outside the perimeter. They support those who will honestly confront injustice past and present. And, they encourage everyone to join together to accomplish shared ideals. As hard as the process is, heritage leaders recognize that it is a journey toward authenticity and fullness.

References

Bales, K. (2016). *Blood and earth: Modern slavery, ecocide, and the secret to saving the world.* New York, NY: Spiegal & Grau.

Banaji, M.R., & Greenwald, A.G. (2013). *Blindspot: Hidden biases of good people.* New York, NY: Delacorte Press.

Fine, C. (2008). *A mind of its own: How your brain distorts and deceives.* New York, NY: W.W. Norton.

Jaynes, J. (1976). *The origin of consciousness in the breakdown of the bicameral mind.* Boston, MA: Houghton Mifflin.

Leopold, A. (1949). *A Sand County almanac and sketches here and there—American Museum of Natural History special members' edition.* New York, NY: Random House.

Lowenthal, D. (1998). *The heritage crusade and the spoils of history.* Cambridge, United Kingdom: Cambridge University Press.

Marshall, G. (2015). *Don't even think about it: Why our brains are wired to ignore climate change.* New York, NY: Bloomsbury USA.

Park, C. (2010). Making sense of the meaning literature: An integrative review of meaning-making and its effects on adjustment to stressful life events. *Psychological Bulletin, 136*(2), 257–301.

Patterson, K., Grenny, J., McMillan, R., & Switzler, A. (2011). *Crucial conversations: Tools for talking when stakes are high* (2nd ed.). Columbus, OH: McGraw-Hill.

Sandel, M.J. (2009). *Justice: What's the right thing to do?* New York, NY: Farrar, Straus and Giroux.

Cannon at battlefield. Courtesy of the National Park Service.

3

Invoking the Spirit of History on the Journey through Hallowed Ground

JAMES A. PERCOCO

THE PLACE

It is mid-November. Thanksgiving break is a week away. Students and teachers are longing for the days off. The first quarter has come to a rapid close and everyone at school is now in "full year" mode. I have dozed off on the charter bus, having met it at 6:00 a.m. with my thirty students; now ninety minutes later, as I look through the haze of my sleep laden eyes, I can see we are approaching our destination, Gettysburg National Military Park. I roust myself and look down the aisle. Students doze, earbuds dangling from many heads. Pillows are propped along breath-steamed windows with heads tilted against them. Blankets cover many of those snoozing. Adult chaperones, too, are sleeping. I sidle up to the driver, giving him directions as we make our way through the south end of the battlefield, through the town of Gettysburg, and head west on the Chambersburg Pike. As we near the crest of McPherson Ridge, I direct the driver to pull off to the right, into the pull-off adjacent to the statues of the Battle of Gettysburg day-one heroes, General John Buford and General John Reynolds. I pick up the microphone and announce that we have arrived. A collective groan emanates from the passengers.

I pull on my wool Civil War colonel's jacket, borrowed from the prop shop at Arlington House National Historic Site, and head down the steps to be greeted by the crisp autumn air. It awakens me further. The students and chaperones descend the same steps, many of the students wearing an assortment of hoodies, and then cluster together for warmth. Vapor clouds form as they yawn and mumble good mornings. The driver cuts off the engine. Traffic in both directions is steady, including noisy tractor trailers, their diesel engines punctuating the morning tranquility.

Now my head is clear. The fog of sleep has dissipated and I am ready for the day. I guide everyone toward the statue of Buford. As I wait for the noisy traffic to subside, I reach into my pocket and pull out a set of Tibetan bells. The students see them and know what is coming. I ask everyone to face west as the sun begins to rise. After a few remarks about where we are standing and why, I turn to face the assembly, bring the bells to front and center, and invoke my classroom mantra. Some of the students chime in: "It's that time in moment and space when we bring the bells together and invoke the spirit of history." The bells meet each other and there is a gentle *cling*, which in the silence has a special kind of reverberance. The spirits and energy of Gettysburg, and all that it means in American memory, rise up to greet us. The sky is clear. The sun is now up and warming us. I know it is going to be a good day!

This field trip took place near the northern end of Gettysburg National Military Park at the northern terminus of the Journey Through Hallowed Ground (JTHG), a 180-mile National Heritage Area. It is a corridor which runs from Gettysburg south to Thomas Jefferson's mountaintop home, Monticello, in Charlottesville, Virginia. The Journey cuts through four states—Pennsylvania, Maryland, West Virginia, and Virginia—yet you can make the drive from Gettysburg to Monticello in just over two hours. Along this swath of land are nine presidential sites, including the homes of Thomas Jefferson, James Madison, James Monroe, and Dwight D. Eisenhower. It is the land of hundreds of Civil War battle sites, many of them small skirmishes, others carrying national significance, including Antietam, Spotsylvania, the Wilderness, and Gettysburg. There are places where America's first residents, American Indians of the Susquehannock, Piscataway, and Iroquois tribes, left their footprints. There are sites from the American Revolution, the War of 1812, the Underground Railroad, and Harpers Ferry, where John Brown and his band of raiders set into motion the Civil War. Within the Journey's boundaries are thirteen national park units, numerous scenic rivers, roads, farms, and small towns, all of which echo and hearken to a time before the internet, smartphones, and iPads.

THE MISSION

The JTHG Partnership, an affiliation of organizations, seeks to collaboratively preserve the beauty and historical integrity of this seminal parcel of America. They call it "Where America Happened," and they include education—particularly of the young—as a central portion of their mission. According to Pulitzer Prize–winning author Geraldine Brooks (2006),

> this remarkable region tells the story of how and where America happened. It has the power to transport us on the greatest journey of all: the journey of empathy and imagination into the lives of the people—famous and unknown, humble, and distinguished—who shaped this country and made us who we are. (p. 2)

The president emeritus of the JTHG, Cate Magennis Wyatt, believes that "the Journey is as long as the Grand Canyon and in terms of history, every bit as deep" (2015). So, too, are educational materials developed by the many partners of the Journey. The educational mission has been a priority of Wyatt's since the beginning, and she crafted a strong alliance among many partners not only to raise awareness about the region with the general public, but also among students of all ages.

The National Park Service's (NPS's) centennial goals, like those of the Journey, are to connect with and create the next generation of park visitors, supporters, and advocates. What is so wonderful about the Park Service's centennial is that the NPS is embracing all kinds of parks, including state and local parks, as part of its celebration. The Journey, and all that has developed under Wyatt's leadership, has a huge role to play in the educational purpose of parks and how parks can be incorporated in the classroom.

THE PROJECT

This initiative to incorporate parks into classrooms has gained traction nationwide, including at the Journey. One such engaging project immerses middle-school students in service learning. The students become the historians, creating vodcasts or mini-movies for historic sites as part of an initiative called *Of the Student, By the Student, For the Student*. The work that students create ultimately becomes part of the official interpretive material of places like Manassas National Battlefield, Fredericksburg and Spotsylvania National Military Park, and

Monticello, among dozens of other places located inside the Journey. Shortly after the JTHG Partnership was formed, seeking a project that would engage middle schools more effectively, the Journey leveraged this model and the outcomes have been strong. This project type is resoundingly endorsed by school administrators, state board of education specialists, and parents alike.

In my role as a history teacher in Springfield, Virginia, I worked with Wyatt to implement a similarly engaging and effective project. When we met, Wyatt and I brainstormed some ideas, ultimately deciding to have students adopt some of the locations in the Journey and create a visual product related to their visit. The projects directly linked to historic preservation, in that each project needed to address how its site was threatened by encroaching development. Students were expected to include a detailed connection of the site to the National Historic Preservation Act of 1966. To kick the project off, Wyatt came to visit my students, speaking with them about the Journey. Her visit did not disappoint. She brought a *National Geographic* map and coffee-table book for each student, and she discussed the practical relevance of the project students were about to undertake.

After Wyatt's initial visit to our classroom, all of my students visited the Journey's headquarters in the historic town of Waterford, Virginia. To select their project sites, students drew sites at random from the Civil War–era, tricornered hat I kept in my classroom closet. All the sites were within a ninety-minute drive of the school. Students called ahead to make their own arrangements with site staff, as this visit needed to be more than a photo opportunity. Some went to Monocacy National Battlefield, the site of an important 1864 Civil War engagement that blunted a Confederate attack on Washington, DC. One student went to National Trust for Historic Preservation's Oatlands Historic House & Gardens. Still another visited Dodana Manor, the home of George C. Marshall, one of those important, often overshadowed figures of American history. The students had two months to complete the assignment, which included framing their visit and final visual results within an historical context, generating preservation recommendations that reflected the National Historic Preservation Act of 1966, and demonstrating their local site's connection to the national narrative of American history. When they were finished, Wyatt and other education staffers of the JTHG came to school to watch the students present their impressive work.

THE IMPLICATIONS

A project such as the one described above is feasible to implement in any classroom, in any part of the country. Every community has a history. Every community has a story and those stories are invariably connected to that national narrative of American history. For example, civil rights sites in the South are not only found in places like Selma, Montgomery, and Birmingham, Alabama. African American communities throughout the South galvanized into action during this period to gain access to public libraries and all schools. Sites in southern Virginia address African American agency during the massive resistance to *Brown v. Board of Education of Topeka* (1954) school integration ruling. Harnessing the local resources in these communities could certainly lead to similarly place-based learning for students.

Many of the sites located within the Journey have a reservoir of educational activities, lesson plans, and strategies for teachers to use easily within a K–12 curriculum. The Civil War Trust, one of the Journey's partners, has countless free digital educational tools to bring places such as Gettysburg to your classroom in Nevada—providing much more interaction for your students than a virtual field trip! Similar partnerships across the nation between NPS units and their surrounding communities could make this model of project learning possible. Even without existing partnerships, teachers could reach out to their nearest national park unit to hear more about existing educational curricula and opportunities to connect to the classroom.

On the cusp of the 2016 National Park Service Centennial Celebration, Jon Jarvis, the former director of the National Park Service said, "Our first century was about bringing people to the parks, but the next century will be about bringing parks to the people." There are so many places and stories across the United States where the educational work of the Journey could be replicated. Skillful use of the Journey Through Hallowed Ground resources will go a long way in making history come alive while promoting citizenship and civics at the same time.

FURTHER READING

Having worked directly from the classroom with the Journey Through Hallowed Ground, I was asked by them to write a book about their exceptional resources. In early 2017, *Take the Journey: Teaching*

American History Through Place-Based Learning was published. On its pages, you can learn how to enliven your travel or visitation experience by tapping into the educational work that they have done. The material in the book includes lesson plans and primary documents that can be used by teachers, or even families, who want to engage in heritage tourism on the weekends or during vacation. Everything in the book is designed to improve citizenship with a keen eye toward civic engagement.

I also recommend David Edwin Lillard's *The Journey Through Hallowed Ground: The Official Guide to Where America Happened*, Deborah Lee's *Honoring Their Past: African American Contributions Along the Journey Through Hallowed Ground*, and National Geographic's *Journey Through Hallowed Ground: Birthplace of the American Ideal* by Andrew Cockburn.

Acknowledgments

The author would like to thank the publishing team at Stenhouse Publishers for permission to use material from his recent book, *Take the Journey: Teaching American History Through Place-Based Learning*, in this chapter.

References

Brooks, G. (2006). *March*. New York, NY: Penguin.
Brown v. Board of Education of Topeka, 347 U.S. 483 (1954).
National Historic Preservation Act of 1966, 16 U.S.C. § 470 (1966).
Wyatt, C. M. (2015, June). Remarks from the Civil War Trust Annual Conference, Richmond, VA. Retrieved from http://www.bguthriephotos.com/graphlib.nsf/keys/2015_VA_CWT_Dinner_150606

Buffalo in Glacier National Park. Courtesy of Donal Carbaugh.

4

Two Different Ways of Knowing the Glacier Area

DONAL CARBAUGH

National parks are known typically as beautiful and sublime places. A quick glance down a list of park names—Acadia, Arches, Bryce, Glacier, Grand Canyon, Smoky Mountains, Yellowstone, Yosemite, Zion—and we easily see in our mind's eye scenic views which create awe in our being. Somehow, we feel the best of our landscapes has been set aside for us. Public figures throughout history have gone so far as to declare the national parks as America's best idea (Bryce, 1913; Burns, 2009; Stegner, 1983). Those so inclined might even see our parks, as John Muir (1901) did, as the "varied expressions of God's love" (p. 556). Indeed, the emphatic declaration of nature's wonders surrounds us in these places and at times fills us with a sense of our best, of awe and inspiration. So how do we *understand* and *know* these places?

Over the years, I have had the great fortune of visiting many of our parks and learning from them. In the early 1980s, I guided tours and served as an oral communication specialist in Glacier and North Cascades National Parks, respectively. Since then, my career in ethnography of communication and cultural discourse analysis has led me to many other parks and people who have shed light on what it means to understand and know these places. Today, my study centers on the ways that we *learn* about a place and how these means of learning impact how we choose to *be* in that place. In this chapter, I will highlight some features of a knowledge system Blackfeet (also known as Amskapi Piikuni) people have taught me in an effort to illuminate *ways of knowing*, *learning*, and *being* that we do not often intimately encounter in our national parks.

THE TWO WAYS OF KNOWING: BACKGROUND

Before we dive into our example, however, we need to differentiate two key ways of knowing: my Blackfeet teachers have called these *contemporary* and *traditional*. As a guide, I found it relatively easy to communicate the "science" of the park—a more contemporary way of knowing. Much more difficult for me over the years was learning other views people had about the park, especially the views my Native American educators called traditional.

National parks are something people have made. Governments have designated them and they have specific boundaries and specific codes of conduct—more *contemporary* ways of *being in place*. The ways of being in a national park, within its variable landscape, need to be taught and learned—fires here not there, no fishing in that lake, dogs not allowed in the backcountry, no entry in that valley. It is all a matter of education, and contemporary education at that. Such ways of being in place are based upon what we each understand about place, nature, land, animals, people, and history. In order to help visitors understand and conduct themselves properly, the National Park Service relies upon policy, interpretation, and other knowledgeable visitors. With this shared knowledge, people more reliably do what best serves the place, wildlife, waters, wind, and our common good.

Our contemporary knowledge about these parks and responsibly being within them is relatively new in the overall picture. There have been other, long-standing *traditional* knowledges active in the areas now designated as national parks. Bruce Weaver (1996) discussed "the mountain people" previously living in areas within what is now the Great Smoky Mountain National Park. Upon the creation of the park, the people's long-standing domestic culture clashed with the park's ways, which resulted in "moving hundreds of people out of their homes" (Weaver, 1996, p. 173). In fact, many indigenous peoples have been similarly dispossessed of their homeland by government decree so a park could be created (Spence, 2000). The First Peoples of our continent lived for centuries in areas that are now parks in ways and on lands that were—and are—known deeply to them. These places are familiar and inextricably entwined with their traditional ways, a body of ancestral wisdom. This traditional understanding bears its own guidelines for being in place, separate from the guidelines espoused by the creators of contemporary national parks. Through some conversations with members of the Blackfeet tribe, I hope to illustrate these powerful, lesser known interactions with place.

DEEP LISTENING: A TRADITIONAL WAY OF LEARNING

Deep listening is a way of learning found in the Blackfeet tribe. The Blackfeet are a people whose ancestral knowledge is inextricably entwined with their place. They have lived since the beginning in and around what is now called Glacier National Park. From the Sweet Grass Hills to the Writing-on-Stone Provincial Park to the Badger-Two Medicine area to the Inside Lakes, Blackfeet have drawn lessons, learnings, livelihood, and inspiration from their homeland. Listening to their traditional sense of this place can help educate us when we visit or live in those locations. I did not know about *deep listening* when I was first introduced to the Glacier area, but after being exposed to this method by Blackfeet traditionalists, and eventually practicing it, I have learned in ways not available to me before. By listening deeply, we—those seeking to learn—can find knowledge about unfamiliar places; we can learn ways of being within that landscape that sometimes escape notice or official policy.

One traditional Blackfeet teacher, Rising Wolf, mentioned deep listening when we spoke about ways of knowing while I was teaching at the University of Montana. Rising Wolf and I were discussing different types of education, and he talked about the ways he learned about who he was and about his surroundings in the Glacier area (personal communication, July 13, 1989). The excerpts from transcripts, here and below, are produced in order to capture vocal qualities in how the speaker produced them. For example, pauses occur at the end of each line, indentations show parallel structures, and larger spacing between lines shows how some verses come together as stanzas.

RISING WOLF ON DIFFERENT WAYS OF KNOWING

1. There's two different ways probably, you can go at it.
2.   One would be traditional, because I was raised by a traditional culture and language . . .
3.   The other would be the more contemporary way, the assimilation, Christianity part of it . . .
4. I was raised within what I would consider a time capsule, because, uh,
5. The area where I live, where the Blackfeet are, are on this reservation.

6.   You were able to communicate spiritually and physically,

7.   You were able to function in two different types of atmosphere.

8.   You were able to travel in time, to a point to where

9.   Your education is not just coming from the secondary type of approach but, uh,

10.   from firsthand knowledge that was handed down through the centuries.

11. And I guess that's where you could probably start it, you know, and going off that as a base,

12. Because everything else from there on up to today is based on that.

These twelve lines speak deeply, and just like the places referenced in them, require careful attention. The first three lines of this quote frame two different ways of living, one being *traditional* and the other *contemporary*.

The traditional way is a kind of communal life that involves, for Rising Wolf, speaking his native Blackfoot language. The language captures—through its use—specific messages about the land and particular ways of living that do not readily translate. Places are identified, stories are told, and sacred ceremonies are learned through the use of this language. This way of identifying where one is, how one acts (and should act), and how one feels carries forward a traditional culture, a deep and ancient knowledge. This traditional way of the Blackfeet predates the current national park by many centuries. The traditional ways are also still embraced by many, like Rising Wolf, who use these ways to navigate the contemporary world.

The contemporary world involves pressures to assimilate, which puts pressure on moving away from one's traditional ways. Not long ago, especially from 1887, Christian missionaries' efforts forced attendance in boarding schools where they cut students' hair, punished them for speaking their native language, and installed programs designed to transform tribal members into farmers or ranchers. The government also implemented land transfers which shrank borders for the reservations, especially between 1855 and 1910, drawing invisible lines on a seamless landscape and transforming some reservation lands into parks. This contemporary world brought diseases like smallpox, land grabs, and terrible misunderstandings resulting in loss of territory and life, the latter most notably in the Baker Massacre. These events are parts of the place, albeit difficult parts of a landscape or culture-scape now resting before us. But these events also are ones we, as Rising Wolf, benefit from learning. Rising Wolf is able to navigate through difficult contemporary pressures by recourse to his traditional ways. In other words,

Rising Wolf's historical ways equip him and many others with knowledge he uses to address his present, contemporary circumstances. There is much to learn from this, as he tells us.

THE LANDSCAPE AS A TEACHER: CONSTRUCTING
FIRSTHAND KNOWLEDGE

If we carefully consider his lines (4–10 above), we find that Rising Wolf considers the landscape of Blackfeet Reservation, writ large, to be a kind of "time capsule" (line 4), a deep preserve of traditional ways. Imagine unveiling a world—a kind of symbolic archeology—that one never knew existed, a portal to personal as well as cultural restoration that helps us live better today! That sort of way and the knowledge it makes available to us is indeed here, in this very place. Rising Wolf draws our attention to the traditional ability to communicate both spiritually and physically to bring forth ancient wisdom or "firsthand knowledge" (line 10). This sort of knowledge is intimately interwoven throughout the landscape, into its animals, the waters, and winds. One can see, hear, and feel not only the physical presence of such "objects" or "things," but along with it, more deeply, spirituality, an enlivened spirited co-presence. The inseparable union of the material and physical realms, people and animals, natural and spiritual worlds, is part of this "firsthand knowledge." As recorded in the Blackfeet Creation story, in the early days all the living things could talk and share their special powers with each other.

"Firsthand knowledge" is difficult to comprehend if it is not within one's own educational system or history. It is also nearly impossible to convey, briefly, in mere written words. Another Blackfeet teacher, Two Bears, accompanied me over the years as a guide to many places in and around Glacier National Park. He would drive us to a special place, pull over, and we would hike for a little while. Then he would stop, energized by the world around him. Pausing, we would scan the horizons, gaze at the tall grasses and wildflowers, smell the air, feel the breeze, be warmed by the sun, watch a raven, hawk, or ground squirrel for a few minutes, and continue onward.

The first time I embarked on an adventure with Two Bears, we were walking through a wide expanse of a meadow, small buttes surrounding us. He was walking in front of me, then stopped. I waited in his tracks for a long time. Not knowing what was going on, and feeling the pause was well beyond what I considered normal, I started to worry a little bit about Two Bears's well-being. Was he okay? After some more time passed, my worry began to border on alarm. Then, Two Bears started walking again. In time, I grew more familiar with this sort of outing. After some decades and much discussion with Two Bears, Rising Wolf, and many others, I began to learn deeply from these elongated pauses.

One time, after stopping and gazing lengthily and longingly over the plains with the fantastic rocky peaks above, Two Bears said a hardly audible "ah." Elsewhere, after hearing the rippling stream and listening to a breeze through the cottonwoods, cooled by its presence, with leaves and seeds flowing around us, he said simply, "A good place." Then, we moved on. Eventually, I began to realize how deeply this routine went, bringing with it a deep capacity to listen, reflect, and learn from my Blackfeet teachers. Over the years, I realized many such places spoke of traditional activities, such as encampments, ceremonies, family histories, and ancestral activities. Sometimes there was physical evidence at a place, like a tipi ring or buffalo bones; at others, not. Those so educated could hear at each place something significant, often a moral lesson to be learned not only about history, but about living today in these historically based ways. Historical knowledge was speaking in the present, perhaps immediately, about present concerns. There was the possibility to learn as the wind blew, as a raven cawed, or as the water moved through the ravine. When so, the natured and cultured place spoke not only of its history but to our present circumstances. It was worthy of close and careful attention. There is much here to learn, and this learning can go deep.

This form of deep listening activates a traditional way of being. With it, Blackfeet can travel today through the historical form, connect with a past, and learn from it about our present and ways of living well within it. It is important to emphasize that this traditional way of learning is not a simple carrier of traditional knowledge, although it can be that, but is a profoundly different way of learning and knowing. Rising Wolf carefully constructs his comments to draw this to our attention. He contrasts "two different types of atmosphere." The one is in the form of a "time capsule" that is on the reservation, housed in its places; this complex set of agents speaks, and as it does, so provides "firsthand knowledge." One experiences and learns accordingly by being on the reservation, in the meadow, on the prairie, by the stream, in the shadow of the great Rocky Mountains.

Rising Wolf elaborated on this form of learning (personal communication, July 13, 1987). He wanted me to know that the material part of the landscape should not be separated from the spiritual part of the landscape. Traditional ways are spiritual ways and yet these are based in the material reality around us. Both dimensions are always already there for us to see, feel, explore, and learn from. This part of knowledge is too often silenced today, he said. This can be challenging for us to understand, confusing to us if we forget. For that reason, ceremonies and rituals that keep both in view together are quite helpful.

RISING WOLF ON NATURE'S OFFERINGS

71. When you're tryin' to communicate with what nature's tryin' to offer you
72.   around you
73. In our prayers
74.   we ask the water
75.   we ask the fire
76.   we ask the air and
77.   we ask the earth to help us
78. We go from the smudge
79.   which is the smoke that goes and carries our prayers to the spiritual world
80. We go to there
81.   we ask for the knowledge of the universe
82.   we ask for the help of mother earth
83.   for the food that she gives us
84.   we give thanks and
85.   ask for more help
86.   we ask the water for everything that is given us
87.   we thank it
88. And in this way
89.   in this direction
90.   we try to do that every day
91.   every day, I mean
92.   in the morning when the sun rises
93.   we pray to the sun for lettin' us
94.   thank it for lettin' us see it one more time
95.   and when it sets and the moon rises
96.   we thank the moon in the same way
97.   for lettin' us see it one more time

Rising Wolf admits the difficulty of "tryin' to communicate with what nature" offers (line 71). Certainly, all of us have grappled with that sort of question. We may have lost our way, unable to hear the spirited-material world around us. Yet somehow, when we stand atop Cadillac Mountain in Acadia, perch on a side of the Grand Canyon, or situate ourselves among the towering walls of Zion, our senses can be opened to learning anew. The world speaks through many dimensions and media. How can we understand what nature has offered us? What knowledge is available to us when we are in these places?

Rising Wolf's response activates the careful attentiveness to nature's offerings (lines 71–97), and his activity is a kind of deep meditation in a sequential form. It is an integrative knowledge in the realm of the material-spiritual. He guides us through that integrative form, in short, as he structures his remarks: "we ask," "we go," and "we thank."

As he says (lines 74–77), we ask water, fire, air, and earth for help and for inspiration. We are reminded (lines 78–79) that appreciating the spiritual dimensions of our material landscapes is just like acknowledging that smoke draws from a fire which ignites material things; smoke, like a material thing, is at first visible, but then can move in ways invisible to the eyes or our other senses. This symbolic form of progression indicates that spirituality is always there, even if not initially seen, or if once seen, is quickly forgotten by us. This reminder is partly the reason for reflective and meditative acts such as "the smudge" (line 78). When we "go to there" and are attuned to the world in these ways, we can possibly learn "knowledge of the universe . . . [from] mother earth." This can offer spiritual and material food for thought (lines 80–83).

Because we can learn in these ways, humbly, from the remarkable presence and parcels of the earth around us, we can (and should) "give thanks" (line 84). Rising Wolf then introduces a cycle of such potential learning, going to "water" then implicitly to fire, air, and earth. There is much to learn in varieties of things both materially and physically, right there, before us. Much is given to us, and because of this, Rising Wolf says, "we thank it" (line 87).

In the final stanza (lines 88–97), he reminds us of "this [traditional] way . . . this direction." It is crucial to practice this form of living and to learn through it every day, repeatedly throughout the day. There is so much we are part and parcel of. Be attentive and attuned to it. We have so much to be thankful for, the ability to be with this earth of ours yet one more time. "The knowledge of the universe" comes to us in varieties of ways, through various expressive media, including wind, water, and the rest. We struggle to learn from this sort of knowledge of the universe, but learn from it we should, as the potential for such learning is great. We must neither forget this nor neglect to act upon it.

TRADITIONAL AND CONTEMPORARY KNOWING: IN OUR LIVES AND IN OUR PARKS

This practice of listening is something of a traditional way for some Blackfeet, especially for elders who live in the traditional way. Yet, as introduced by Rising Wolf, this way can be difficult to practice, live, and realize. There is much in its way; the winds are blowing strong across our land from that other "atmosphere." The *other way* is, at times, a grand storm of difficult-to-navigate dust, a "contemporary world" ushering in a different "atmosphere." For Rising Wolf, Two Bears, and others, this contemporary way is "secondary" to the traditional way. And that perspective on it can be difficult to maintain, especially for younger generations.

By juxtaposition, the contemporary way brings with it a predominant way of learning and specific forms of knowledge. This is sometimes called "Whiteman's education." This draws attention to the sort of thing one might learn simply from writings or books or whiteboards, in a classroom. This sort of learning is indeed considered important and valuable. Yet at the same time it is distinguished as different from the traditional way, less connected to the roots of a Blackfeet tradition and its places. It is typically focused on the physical without the spiritual or fixed only on one side of the coin of knowledge. As we listen to the final line offered by Rising Wolf, in "different ways," he says, "everything . . . is based on [the traditional way]" (line 12). This reminder of the bases of traditional ways and their ancient wisdom brings to the fore Blackfeet knowledge in its places, of where Blackfeet people are, a centuries-old homeland, what has occurred there, what should occur there, and how people knowing this tradition feel about those places, activities, and inhabitants. The latter includes not only the literature written by the "two-leggeds," but also the teachings of the water, wind, and "four-leggeds." All that is here and there speaks to us; all that was here and there deserves our careful attention.

Other Blackfeet have spoken about the importance of listening in these ways and learning from the world around us. Joe McKay, a member of the Blackfeet Tribal Business Council, was speaking on September 2, 2015, at a public meeting on historic preservation in Choteau, Montana, to a native and nonnative audience. He stressed that anyone can learn something in the traditional way, but it takes the sort of special attentiveness Rising Wolf has emphasized above. As Mr. McKay put it,

> "I was taught that
> you don't have to be an Indian
>   you don't have to be Blackfoot
>   you don't have to be Amskapi Piikuni.
> You just have to believe.
> And if you go there
>   and open your heart
>   and your mind
>   and free your spirit
> Then you'll experience what we experience.
> Until you've been there and experienced it
>   it's really difficult to describe." (McKay, 2015)

Joe McKay reminds us that "you just have to believe . . . then you'll experience what we experience." It may not be quite that simple to so believe, for the uninitiated, but if we work at doing so, and listen deeply, there are worlds of things we can learn. There is depth of knowledge in our places from which we can learn, difficult as that may be.

Joe McKay, along with Rising Wolf and Two Bears, reminds us of the need to pause, look, and listen to where we are, what has transpired there, and what we can learn from it. He admonishes all of us: "Open your heart, and your mind, and free your spirit." If we are somehow able to do this, we can experience the place, the Glacier area and other areas, in old yet new ways. This is the work of an ancient tradition, but perhaps it is an underused human capacity today. If exercised in this way, the "knowledge of the universe" stands before us, right there in our parks!

Can we learn in all the ways our lands make available to us, in our places? Can we indeed be educated by them? Yes, we have much to learn, including from an expanded way of learning itself. If cultivated, this will allow us, respectfully, by honoring what has come before, to better navigate the diverse atmospheres surrounding us in our places by our different ways, and to learn from yesterday, while forming a better today and tomorrow.

Acknowledgments

This chapter is dedicated to Bettina Burke, Will Henderson, Darrell Kipp, Two Bears, Curly Bear Wagner, Rising Wolf, Jack Gladstone, Bonnie Heavy Runner Craig, Leon Rattler, Woody Kipp, Joe McKay, and others who offered their time and goodwill to me over the years: all thanks and no blame.

References

Bryce, J. (1913). National parks: The need of the future. In J. Bryce (Ed.), *University and historical addresses* (pp. 389–406). New York, NY: MacMillan.

Burns, K. (Producer). (2009). The national parks: America's best idea [Television series]. Washington, DC: Florentine Films & WETA.

Muir, J. (1901). The fountains and streams of the Yosemite National Park. *The Atlantic Monthly, 87*(519), 556–565.

Sakariassen, A. (2015, September 10). Badger-Two Medicine: A united front. *Missoula Independent*. Retrieved from https://missoulanews.bigskypress .com/missoula/badger-two-medicine/Content?oid=2444173

Spence, M.D. (2000). *Dispossessing the wilderness: Indian removal and the making of our national parks.* New York, NY: Oxford University Press.

Stegner, W. (1983). The best idea we ever had: An overview. *Wilderness, 46*(160), 4–13.

Weaver, B.J. (1996). "What to do with the mountain people?": The darker side of the successful campaign to establish the Great Smoky Mountain National Park. In J. Cantrill & C. Oravec (Eds.), *The symbolic earth: Discourse and our creation of the environment* (pp. 151–175). Lexington: University Press of Kentucky.

Student making observations through a microscope at Canyonlands National Park. Photo by Kirsten Kearse; courtesy of the National Park Service.

Feedback Loops

Systems and Science Learning

This section explores the ecosystem of learning, through learning about ecosystems, science, and research in national parks. From climate change to citizen science curriculum, the authors partnered with national park sites across the country to better understand how research can inform place-based learning and how learning can inform place-based research. Just like dynamic systems in nature, the ecosystem of learning is characterized by feedback loops.

Davis and Thompson remind us that America's national parks are changing in their chapter, "Learning about Climate Change in Our National Parks." Climate change has been recognized as possibly the greatest challenge ever faced by the country's land management agencies. National Park Service (NPS) units throughout the United States are already feeling and seeing impacts to natural and cultural resources. The authors explain the results of more than three thousand surveys and 350 interviews with visitors to national parks in an effort to understand what visitors already know about climate change and its impact on natural and cultural resources. Essentially, visitors feel connected to the park (even on the first visit), are concerned about climate change, and expressed a strong desire to learn more about climate change while visiting national parks.

The next chapter, "Place-Based Education at Teton Science Schools: Inspiring Curiosity, Engagement, and Leadership in National Parks and Beyond," focuses on programming strategies and examples to engage visitors in place-based learning. Krasnow, McClennen, Kern, Leary, and Peck introduce us to the nationally recognized Teton Science Schools (TSS), which was established in 1967 and has a permanent campus within the boundaries of Grand Teton National Park. Over the past twenty years, TSS has expanded place-based curriculum for school groups, community members, and park visitors of all ages, serving more than fifteen thousand learners annually. This chapter presents snapshots from several of TSS's successful programs. With a mission for inclusive and transformative learning, they have developed programs that help to understand the ecosystem and empower citizen scientists. For example, they have facilitated everything from an invasive fish study conducted by local fifth-grade students to an analysis of ecotourism impacts through wildlife expeditions.

TSS is not the only NPS partner facilitating citizen science opportunities; currently there are more than 430 active citizen science projects on US public lands (see www.citizenscience.gov). Houseal, an expert in science education and curriculum development, proposes a framework to deepen learner involvement in citizen science initiatives. Her chapter, "Three-Dimensional Learning: 'Upping the Game' in Citizen Science Projects," explains the Framework for K–12 Science Education published by the National Research Council and how it was developed to guide science standards for formal educational settings. This framework proposes that learning science is dependent on three dimensions: science content, scientific processes, and overarching ideas that span the discipline. When these three are integrated, the scientific enterprise can be understood in a more realistic and authentic manner. This framework, when integrated in citizen science project planning, has the potential to increase the richness of participant involvement in citizen science projects, especially in national park settings.

The final chapter in this section, "Mentoring Mountain Raingers: Beyond Basic Hydrological Field Research in the Great Smoky Mountains," describes the day-to-day experience for undergraduate students conducting high-elevation rain gauge research. Miller walks the reader through the practical challenges and long-term benefits of designing an in-depth place-based research program with college students. From building and installing the rain gauges to encounters with wildlife, this was an in-depth learning experience in Great Smoky Mountains National Park for the student researchers. Miller explains that it wasn't only about collecting data; there are many learning, professional, and personal benefits for students and university research faculty involved in a long-term field project, like this Duke University and the University of North Carolina Asheville partnership.

Interpreter arm holding picture of glacier recession. Courtesy of Kaitie Huss.

5

Learning about Climate Change in Our National Parks

SHAWN DAVIS AND JESSICA L. THOMPSON

We were in Kenai Fjords National Park in Alaska. We signed up for the ranger-led glacier walk. As we walked the paved trail to Exit Glacier, we noticed a number of simple signs with seemingly random years (e.g., 1917, 1926, 1951, 1961, 1997, and 2005) engraved on them. Curiosity prompted us to ask the ranger about these signs; immediately the interpretive ranger had us hooked. Ready, with pictures in hand, he showed us graphs and photos of the glacier's shrinking path. In the hands of a master interpreter, we were walking the path of a warming climate. Exit Glacier is one of the most accessible glaciers in Alaska. Its retreat has been monitored for more than a century, and the rapid recession highlights the effects of climate change and how it affects Alaska's coastal glaciers. In the early 1800s the glacier retreated an average of 3 feet per year, but by 1899 it was shrinking at an average of 43 feet a year (Huse, 2012). In 2014, Exit Glacier retreated approximately 187 feet-in just one year. In 2015, President Barack Obama visited Exit Glacier during his historic trip to Alaska. Hiking to Exit Glacier is one of the most popular visitor activities at Kenai Fjords National Park, but that experience is becoming more difficult every year. The trail has nearly doubled in length from its inception; we are chasing a continuously retreating glacier.

In 2010 we had an opportunity to investigate visitors' perceptions of climate change in national parks across the country. We surveyed and interviewed visitors; we worked with national park staff to identify informational and programming needs; and we traveled to some of the most iconic public lands in our nation. We knew that this research would be engaging and awe inspiring, but we had no idea we would learn so much.

On this journey, we hiked to and touched the terminus of Exit Glacier in Kenai Fjords National Park, Alaska. We explored an underwater world at Biscayne National Park in Florida. We were immersed in our nation's history in Washington, DC. We marveled at Mount Rainier and snowshoed through Rocky Mountain National Park. In all, we visited ten national parks, including Mount Rainier, North Cascades, Olympic, Rocky Mountain, Kenai Fjords, Everglades, Biscayne, National Capital Parks East, Harpers Ferry, and Prince William Forest. These parks were chosen in collaboration with the National Park Service (NPS) to represent areas that differed geographically, ecologically, and demographically and were experiencing a range of climate change impacts, as identified by the NPS.

Our investigation included completing (a) a comprehensive literature review of climate change science and research on impacts relevant to the selected national park sites; (b) interviews and surveys with agency managers and frontline staff (35 interviews, 847 surveys); (c) interviews and surveys with park visitors (359 interviews, 3,233 surveys); (d) five regional workshops; and (e) ten site visits and focus groups with agency staff. Throughout these research activities we did not realize the profound effect this project would have on our understanding of the issue and the importance of place-based public engagement and education about the dynamics of climate change impacts. Most importantly, we learned that park visitors were quite possibly the perfect audience to motivate and mobilize to take action on climate change.

INSPIRATION FOR THE PROJECT

In late August 2009, US Senators Mark Udall (CO) and John McCain (AZ) were at Rocky Mountain National Park for an Energy and Natural Resources Subcommittee meeting. This was an official US Senate meeting, but a portion of it was open to the public. We attended and learned about the impacts of climate change on Colorado's parks, including an explanation of the bark beetle infestation in Rocky Mountain National Park's iconic pine forests, invasive cheatgrass spreading to higher elevations, melting rock-glacier ice, disappearing pikas, and the fragmentation of large mountain ecosystems. The next morning the local newspaper quoted Senator McCain: "Every home in America should see what's happening here. . . . Look, I believe climate change is real. Every visit we make, such as we are making here, argues that we need to take action" (Hanel, 2009).

We discussed what we had learned during the field trip and reflected on McCain's idea-what if people could understand climate change impacts through meaningful dialogue, situated in a specific place or landscape where they could discuss the deeper ecological and social interrelationships and impacts? Would people understand climate change better? Would they feel motivated to make changes in their own personal behavior? Would they advocate for climate-friendly policies? Could it make a difference?

Around the same time we were involved with the Association for Conflict Resolution (a national organization of conflict resolution professionals and practitioners), and the leadership team was planning its annual conference in Denver. A colleague called and said, "Hey! Don't you work with the Park Service on projects? Can you get us into the park for a conference field trip?"

Still inspired by Senators McCain and Udall's visit, we organized a field trip, but it was really an experiment. Could we take a group of people to a park and talk about climate change in a productive way? Could we facilitate a conversation that wasn't "doom and gloom" or buried in scientific jargon? What would an inspiring, place-based conversation about climate change look like? We called a couple of natural resource colleagues at Rocky Mountain National Park and set up a half-day guided tour, with a focus on climate change impacts.

On the morning of the excursion, about twenty people with varying interests in climate change loaded into the conference vans. Some of the participants just wanted to go to the entrance gate to get a picture of the sign; others came along because we promised hiking and wildlife viewing. It was an interdisciplinary group with diverse backgrounds and interests. Our guide met us at the Beaver Meadows Visitor Center and began to explain some of the research the natural resource managers conduct at the park, including several of the climate change impacts they are monitoring. We went on a short walk around Sprague Lake where she talked about water, rock-glacier ice, and changing animal habitats. Then we went to Glacier Basin campground where we discussed bark beetle infestation and the subsequent management decisions.

Partway through the Glacier Basin presentation, one of the field trip participants, a middle-aged woman, put her hand on my arm and said, with tear-filled eyes, "I wish I had brought my parents." Pulling me aside, she told me how she camped at this campground every other summer as a child. Her parents were older now, strong conservationists with a strong appreciation for the great outdoors, but an equally strong disbelief in human-caused climate change. She remembered this campground, riding her bike with her cousins, playing cards at the picnic table, and making sandwiches before a hike. She looked around, tears in her eyes. "Now it is gone . . . and my parents don't get it. They need to see this."

It was gone. From the park's beginning through the late 1990s, Glacier Basin campground was covered in thick, tall lodgepole pines. It was a deep forest with small squares for campers to nestle into the mountain woods. It was rustic, private, and most importantly, forested. By 2009, all of the lodgepole pines that were infected with bark beetles had to be cut; they were a liability. The park could not risk them falling on campers or catching fire.

We weren't quite expecting this type of emotional reaction to a discussion about park management policies. We weren't prepared for this type of impact; we merely wanted a McCain-level impact, but we got an intense, emotional, place-based dialogue that engaged all of the tour participants. It was a conversation about science, family, memories, and the legacy of the national parks. We also realized that stories of climate change and landscape changes may require in-house grief counselors at some of America's national parks.

CLIMATE CHANGE IMPACTS AT NATIONAL PARKS

America's national parks are changing. Climate change has been recognized as possibly the greatest challenge ever faced by the country's land management agencies (Delach & Matson, 2010). NPS units throughout the United States are already beginning to see impacts to natural and cultural resources (Bentz et al., 2003; Millar, Westfall, & Delany, 2004; Moritz et al., 2008; Salazar-Halfmoon, 2010) and ongoing research continues to reveal how our changing climate is affecting all public lands across the country. Some of these impacts include sea level rise and ocean acidification in Everglades and Biscayne Bay National Parks (e.g., Wanless & Vlaswinkel, 2005), changes to estuaries and species range shifts in the greater Washington, DC, area (e.g., Gonzalez, Neilson, Lenihan, & Drapek, 2010; La Sorte & Thompson, 2007; McMahon, Parker, & Miller, 2010; Scavia et al., 2002), glacial retreat, drought and species shifts in Kenai Fjords National Park in Alaska (e.g., Adalgeirsdottir, 1997; Berg, 2006; Klein, Berg, & Dial, 2005; Kyle & Brabets,

2001), altered river and marine ecosystems in Olympic, North Cascades, and Mount Rainier National Parks (e.g., Elsner et al., 2009; Huppert, Moore, & Dyson, 2009; Mote, 2003; Mote & Salathé, 2009), and snowmelt, streamflow, and vegetation changes in Rocky Mountain National Park (e.g., Clow, 2010; Hicke, Logan, Powell, & Ojima, 2006; Mote, Hamlet, Clark, & Lettenmaier, 2005; van Mantgem et al., 2009). With impacts all around us, citizens are exposed to many messages about climate change on a daily basis, yet studies show a declining trend in public understanding of human-caused climate change (Stern, 2007). In fact, a recent study found that more than 75 percent of Americans are aware of climate change but less than 50 percent see it as a serious threat (Lee, Markowitz, Howe, Ko, & Leiserowitz, 2015).

Engaging Visitors in Climate Change

Many factors have explicitly challenged the effective communication of climate change science with the public. First, there is an enormous time lag in the change in climate and changes in our social system, coupled with the assumption that the impacts of climate change will most directly affect the developing world (Lorenzoni & Pidgeon, 2006; Moser & Dilling, 2007). Second, there is a widening gap between the public's awareness of what action is needed and what actions are being taken. Without an understanding of what to do, individuals are left feeling overwhelmed and frightened, or blissfully ignore the magnitude of the issue through denial (Moser & Dilling, 2007).

Compounding an individual's feeling of being overwhelmed, blissful ignorance, or outright denial that climate change is happening is a lack of understanding of the science of climate change. When it comes to climate change literacy in the United States, the average American would score 54 percent (i.e., they would fail) on an eighty-one-question test about climate science, climate change impacts, and earth systems (Leiserowitz, Smith, & Marlon, 2010). One issue influencing the lack of climate literacy in the United States is that most people get their information about climate change through a mediated source such as television news, the most popular source (Maibach, Wilson, & Witte, 2010), and when climate change is reported in the news it is often accompanied by images of weather disasters. From earlier research (i.e., Bostrom & Lashof, 2007; Reynolds, Bostrom, Read, & Morgan, 2010), we know that the public understands weather and natural disasters as *acts of god* and fails to see that human actions and lifestyle choices are capable of influencing the pace of climate change. Overcoming this challenge requires that climate change communicators figure out how to translate the complexity and nuances of earth systems for diverse audiences, and then help the audience make connections between human choices and the acceleration of climate change impacts.

Place Attachment to Parks

In visiting national parks across the country, we easily understood the connection visitors felt to these areas. We, too, were captivated by the awe-inspiring peaks of the Rocky Mountains in Colorado, the peaceful and verdant quiet of the Hoh Rainforest in Olympic National Park, the abundant wildlife of the Anhinga Trail in the Everglades, and the landscape-altering power of Exit Glacier in Kenai Fjords National Park. In these areas, it was easy to feel a sense of familiarity with these natural, though somewhat foreign, landscapes. People build emotional relationships with specific landscapes, something for which Altman and Low coined the phrase *place attachment*, or the bonding of people to places (1992). Where we live, work, and, indeed, recreate all have a profound influence on how we define ourselves and our behaviors (Raymond & Brown, 2011). Ordinarily, place attachment takes time to form and is often attributed to places where individuals have spent years visiting and revisiting the same unique landscape. In our research at national parks we found something quite different. Most of the people we surveyed ($n = 3,233$) were visiting the park for the first time and repeat visitors were rare. Still, we saw overwhelming self-reported attachment scores to these areas (figure 5.1). Perhaps this was due to the naturalistic and inspirational landscapes that harkened back to an image of an idyllic, untouched scene from the nation's distant past. This could also be due to the social context; visitors often travel with family and friends, and the norms of the social group can exert influence. Regardless of the cause, these places and the attachments to them are important and often lead to a sense of ownership and protection.

Most visitors form an ideal conception of a park based on their primary experiences there and are reluctant to yield this mental construction of how the park *should* be. As climate change continues to alter our national parks, those with strong attachments to these cherished places must find ways to reconcile these differences. Indeed, if left unresolved, these disturbances to places with strong relational ties can even lead to psychological disorders (Fullilove, 1996). However, if a place of particular importance

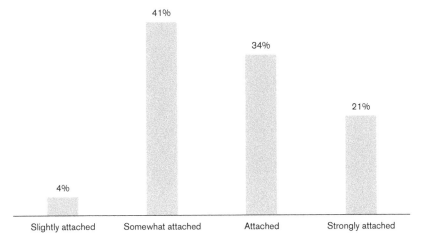

FIGURE 5.1. Percentage of surveyed visitors who felt an attachment to the park (*n* = 3,233).

to a person is threatened, they will participate in actions to protect that special place (Stedman, 2003). Knowing the threat, in this case, is vitally important in deciphering the appropriate mitigating action. In some cases of place disturbance, the threat, and therefore the solutions, are easily discernible. For example, if a housing development is proposed to be built in a beloved neighborhood forested park, the obvious action would be to participate in town hall meetings and raise awareness in the community. In the case of climate change impacts, however, both the cause and solution are not always readily apparent. In these cases, the visual effects, if not already recognizable, should be made so and the connection to a changing climate as the cause should be paramount.

As an example, one of the projects that stemmed from our research was a collaborative repeat photography project with the Southwest Alaska Network (SWAN) and Kenai Fjords National Park named *Making Sense of History: Understanding Landscape Change in Alaska* (see Mullen, Newman, & Thompson, 2014). In this project, visitors are encouraged to take photographs of glaciers at particular photo kiosks. Participants can then upload their photograph to a website where they can compare their photo to other participant and historical photos taken from the same spot throughout time. The temporal scale makes visual changes more obvious. Additionally, these changes are explained as a product of climate change on the website, with links to scientific climate change studies of the area. Finally, mitigating actions are

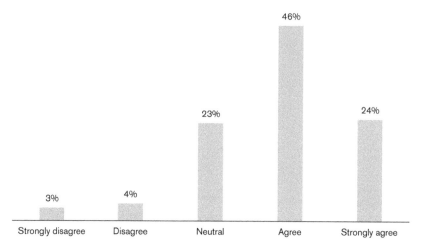

FIGURE 5.2. Percentage of surveyed visitors who agreed that the effects of climate change can already be seen at the park (*n* = 3,233).

suggested, such as switching to a green energy provider, for participants to engage in throughout their daily lives.

Overall, our research has shown a population of visitors who see the impacts of a changing climate in the national parks we have visited (figure 5.2); however, this was not evenly distributed across the different parks we studied (Schweizer, Davis, & Thompson, 2013). In areas such as National Capital Parks East in Washington, DC, the effects and linkages to climate change were somewhat elusive. These included changes to vegetation patterns (e.g., Gonzalez et al., 2010), as well as phenomena such as the urban heat island effect, superheated runoff, and erratic weather and storm intensity. In such cases, connecting the visible phenomena in the parks to climate change is the recommended action. In this particular example, we recommend interpreters highlight the changing phenology of the cherry blossoms and thus the impact to the Annual Cherry Blossom Festival in Washington, DC (Abu-Asab, Peterson, Shetler, & Orli, 2001; Chung, Mack, Yun, & Kim, 2011).

Seeing Is Believing

As one visitor stated in an interview, "[Climate change] didn't mean a whole lot until I'm seeing this stuff. It's happening. I guess I didn't think it was that important before. I now see that it is." The ability to see the effects of climate change in these areas is also vitally important to education. In our research, the ability to recognize the physical effects of climate change was substantially correlated with the visitors' desire to learn more about climate change, greater than any other variable (Schweizer et al., 2013; Thompson, Davis, & Mullen, 2013).

Free-Choice Learning about Climate Change

One of the greatest challenges in education is motivating students to learn about a subject in which they are not interested. In contrast, we love to learn about subjects that we are inherently drawn to. This can be seen in every child who has memorized the names of different dinosaurs or any adult birder who continues to add to their "life list." When we actively seek out information that is of interest to us, we may be engaging in what has been termed *free-choice learning*. Free-choice learning is guided by the desires and motivations of each idiosyncratic learner and therefore exhibits different learning outcomes as varied as the learners themselves (Falk, 2005; Falk & Dierking, 2002). Though we can engage in free-choice learning nearly anywhere, it typically occurs in areas such as national parks, national wildlife refuges, aquariums, zoos, and museums where a highly structured learning atmosphere is absent (Falk, 2005; Falk & Dierking, 2002; Falk & Heimlich, 2009). Kola-Olusanya (2005) notes that free-choice learning venues such as nature centers, parks, and wilderness offer the greatest opportunity for environmental education due to direct experience in pristine environments. According to Falk and Dierking (2002), free-choice learning integrates three factors: place, person, and others-also known as the physical, personal, and social contexts. In other words, this style of learning occurs in particular places where the learner can discuss and form personally relevant knowledge with friends, family, and others.

In studying visitors to national parks in 2011, we came across a surprising finding. As we met and discussed with rangers and staff of the NPS, we heard the common refrain that visitors were not interested in climate change. Visitors came to the parks to see the beautiful vistas and landscapes, they wanted to learn about the cultural and historical contexts, they were interested in the wildlife that inhabits the park, but they were not interested in learning about climate change. In a survey of more than four hundred park service employees, we asked, *"How concerned are your visitors about climate change?"* Respondents were allowed five choices, ranging from *"not concerned"* to *"extremely concerned."* As seen in figure 5.3, the answers to this survey question support

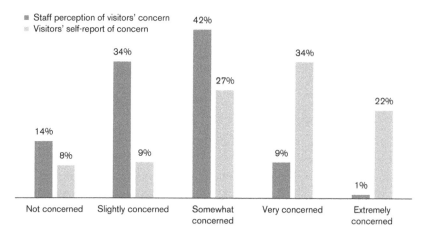

FIGURE 5.3. NPS staff's perception of visitors' concern about climate change compared to the visitors' self-reported concern about climate change (staff *n* = 788; visitors *n* = 3,233).

statistically what we had been hearing anecdotally from the rangers and staff.

To follow up on the staff's perception of their audience, we conducted a separate survey wherein we asked over three thousand national park visitors how concerned they were about climate change. The results show a distinct gap between the NPS staff's perceptions of the visitors and the visitors' self-reported level of concern about climate change.

These results provide us with two main findings. The first, as seen in figure 5.3, is that visitors are concerned about climate change in national parks. This is good news, as this concern also correlates with visitors' desire to learn about climate change. However, the more important finding is that the NPS staff may not know enough about their visitors to create effective educational programs. The first rule of communication is to know your audience so that you may provide them with the most relevant information in the most applicable way. When considering the challenges in communicating about climate change, educators and interpreters, not surprisingly, are apprehensive about discussing the subject. We often heard from rangers we interviewed that they feared backlash from potential climate change deniers who might be present at one of their talks. These fears were expressed by an NPS interpreter, who explained "For me, interpreting climate change impacts is like interpreting the Civil War. I hope I never have to do it because I'm certain that half the group will be neo-Confederate nay-sayers." Recall that we are more apt to learn and remember what we have learned when we are intrinsically interested in the topic. In this case, we can see that visitors are not only concerned about climate change, but they want to learn more about the subject as well. Only 8 percent of the visitors we surveyed did not want to learn about climate change in national parks.

PLACE-BASED CLIMATE CHANGE EDUCATION

Ask anyone to name one animal that is currently under threat due to climate change and you are likely to get the same answer: the polar bear. Gravely threatened by the changing climate, the polar bear has become the poster child for climate change, something Graber (1976) would call a "condensation symbol," encapsulating our understanding and anxiety over our warming planet (Cox, 2009). However, not many of us have seen a polar bear in the wild. Though readily recognizable, the polar bear does not engender a meaningful connection with many park visitors (outside of Alaska), as it falls outside of the scope of reference for many Americans. Many species are currently under threat due to climate change and the national parks give us the ability to see, connect, and empathize with a wide variety of flora and fauna. For example, the American pika, a small relative of the rabbit, inhabits the high peaks in Rocky Mountain National Park. Observing this small creature scurry to and fro among the rock faces is a favorite pastime of many Rocky Mountain visitors. Unable to sustain itself at warmer temperatures, the American pika is currently threatened by climate change. The American pika is a more relevant choice (than the polar bear) for discussions about climate change impacts at Rocky Mountain National Park. Similar place-based options exist for educating about climate change in every park, as the impacts are different in each area. Focusing on local examples is more meaningful, leads to greater empowerment, and helps dispel feelings of helplessness that larger global contexts can induce (O'Neill & Nicholson-Cole, 2009).

Place-based education may be a relatively recent pedagogical term, yet the practice of this particular style of education has existed for more than a century (Woodhouse & Knapp, 2000). Often, the term *place-based education* is confused with experiential education, outdoor education, and environmental education. Though these terms connote similar meanings, place-based education "emerges from the particular attributes of a place" (Woodhouse & Knapp, 2000, p. 4). What is important to realize is that this form of environmental education directly connects learners to the land

and their community, involving both the physical and social fabrics that create a place and make it special. Not only will the focus on local flora, fauna, and ecological processes aid in the formation of place attachments, but they will also engage the learner in issues relevant to their community and local citizenship. As David Orr (1994) notes, education carries responsibility and should be used for sustainable ends, emphasizing the importance of education for real people and real communities. Place-based education produces the social engagement in environmental preservation that many communities currently need (Sobel, 2004).

Although place-based education has been shown to raise academic achievement in many instances (Sobel, 2004), it shows even greater promise in educating about critical social issues, such as climate change and its impacts. Thomashow (2002) states the most effective way to understand and learn about the changes in the environment is by developing an intimacy with the community around you. The more time visitors spend in national parks, the more easily they will learn about the changes taking place there. It is essential that people be encouraged to understand and appreciate local environmental processes before trying to digest the complexity of global climate change and make appropriate behavior changes. Sobel (2004) observed that

> authentic environmental commitment emerges out of firsthand experiences with real places on a small manageable scale (p. 34). . . . What's important is that [people] have an opportunity to bond with the natural world to learn to love it, before being asked to heal its wounds (p. 9).

Seeing the environmental effects caused by climate change through place-based education can also lead to pro-environmental behaviors. In our research, we witnessed a positive correlation between visitors who reported seeing the effects of climate change in the park and their willingness to adopt mitigating behaviors (Schweizer et al., 2013). It seems the more aware visitors were made of the changes, the more environmental actions they wanted to take. Hess, Malilay, and Parkinson (2008) also stress the benefits of localizing climate change messages:

> In particular, a focus on place emphasizes the local nature of both exposures and response, and it brings attention to environmental changes where the motivation to address them is strongest: Emphasizing place highlights climate change's effects where they are most acutely felt, where local strengths are best understood, where place attachment can be leveraged most effectively; where residents will reap the benefits of adaptive measures promoting sustainable and liveable communities. (p. 476)

PLACE-BASED CLIMATE CHANGE ENGAGEMENT

We have thus far discussed how America's national parks are particularly well suited for climate change education. The national parks attract an engaged audience and are physical and social venues that are conducive to learning about a changing environment. However, climate change is a complex and multifaceted subject, which raises the question *What aspect of climate change should be taught?* Discussing the scientific global causes such as the greenhouse effect is far different than discussing local impacts such as salt infiltration due to sea level rise in the Everglades. Though stating local effects is important, our research shows that most visitors are interested in learning about actions they can take to mitigate climate change (Schweizer et al., 2013).

In 2011, Yale Project on Climate Change Communication (YPCCC) released an updated research report that segments the American population into six different audiences who range along a spectrum of concern and engagement on the issue of climate change (Leiserowitz, Maibach, Roser-Renouf, & Smith, 2011). Leiserowitz, Maibach, and Roser-Renouf (2009) describe the different segments as follows:

> The *Alarmed* are fully convinced of the reality and seriousness of climate change and are already taking individual, consumer, and political action to address it. The *Concerned*-the largest of the six Americas-are also convinced that global warming is happening and a serious problem, but have not yet engaged the issue personally. Three other Americas-the *Cautious,* the *Disengaged,* and the *Doubtful*-represent different stages of understanding and acceptance of the problem, and none are actively involved. The final America-the *Dismissive*-are very sure it is not happening and are actively involved as opponents of a national effort to reduce greenhouse gas emissions. (p. 1, italics added)

Each of these segments should ideally have different messages and perhaps messengers; however, with little ability to quickly segment an audience in a national park, the best method may be to aim for a message that reaches the most people. In the 2011 YPCCC study, the researchers found the proportion of the US population in each segment to be as follows: Alarmed (12%), Concerned (27%), Cautious (25%), Disengaged (10%), Doubtful (15%), and Dismissive (10%) (Leiserowitz et al., 2011). However, when we conducted a similar segmentation analysis with national park visitors from our study, we found a significant difference. There was a profound shift toward the Alarmed category among national park visitors, which made it the largest segment of individuals at 29 percent (figure 5.4).

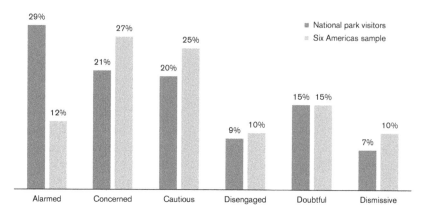

FIGURE 5.4. Comparison of the surveyed park visitors (n = 3,233) to the Six Americas survey respondents (n = 981) and the difference in their concern about climate change.

Of the six segments, the Alarmed segment is the most likely to take action on the issue of climate change through multiple forms, including political activism, consumer activism, energy conservation, and donations (Leiserowitz et al., 2009). Discussing the causes of climate change will not be new information for this group of learners. As this segment is the most likely to take action, messages encouraging climate change mitigation would be best suited for this group. In fact, our study shows that over 90 percent of visitors are willing to change their behaviors within the park to mitigate climate change impacts (Schweizer et al., 2013). The NPS could model behavior change by highlighting its own environmentally conscious policies and practices. For example, Rocky Mountain National Park discussed its shuttle service as a way of cutting carbon emissions in the park and encouraged visitors to use public transportation in the park and at home.

Lessons from the Field

During our research project, we saw several innovative climate change programs at all of the NPS sites we visited. We would like to conclude by introducing the Cascades Climate Challenge program at North Cascades National Park (NCNP) in Washington state. The Cascades Climate Challenge program combined all the key aspects of this chapter: (a) knowledge of audience, (b) place attachment, (c) free-choice learning, and (d) engagement for action in a dynamic curriculum for teenagers. This program was a partnership between the NCNP, the North Cascades Institute, and the

National Park Foundation. The program engaged high school youth leaders from across the Pacific Northwest in a month-long climate camp experience at NCNP. As part of the program, participants helped park scientists collect and analyze climate science data about weather, plants, and aquatic species. They identified examples of climate adaptation and made proposals for park management to mitigate impacts. After the month-long camp ended, the students were required to develop service-learning projects in their home communities and to continue communicating about climate change via presentations to peers and the public.

In terms of place attachment, these immersive, long-term experiences offer youth an unparalleled opportunity to deeply connect to the park. They are seeing the impacts of climate change through their studies with the help of knowledgeable rangers. Many of their service-learning projects are self-directed, allowing the participants choice in what aspect of climate change they wish to study. The entire program is place-based, involving not only the park, but carrying over in meaningful ways to the students' residential communities. Finally, the emphasis on action is paramount. The students do not end the program simply learning the science and impacts of climate change, but they are prompted to engage in service and present the information to their communities, prompting discussion and further action on climate change, arguably the most important issue facing their generation.

References

Abu-Asab, M.S., Peterson, P.M., Shetler, S.G., & Orli, S.S. (2001). Earlier plant flowering in spring as a response to global warming in the Washington DC area. *Biodiversity and Conservation, 10,* 597–612.

Adalgeirsdottir, G. (1997). *Elevation and volume changes on the Harding Icefield, southcentral Alaska.* Unpublished master's thesis. University of Alaska, Fairbanks.

Altman, I., & Low, S.M. (1992). *Place attachment, human behavior and environment.* New York, NY: Plenum Press.

Bentz, B.J., Regniere, J., Fettig, C.J., Hansen, E.M., Hayes, J.L., Hicke, J.A., . . . Seybold, S.J. (2003). Climate change and bark beetles of the Western United States and Canada: Direct and indirect effects. *BioScience, 60*(8), 602–613.

Berg, E.E. (2006, November). *Landscape drying, spruce bark beetles and fire regimes on the Kenai Peninsula, Alaska.* Paper presented at the Third International Fire Ecology and Management Congress, San Diego, CA.

Bostrom, A., & Lashof, D. (2007). Weather or climate change? In S.C. Moser & L. Dilling (Eds.), *Creating a climate for change: Communicating climate change and facilitating social change* (pp. 31–43). New York, NY: Cambridge University Press.

Chung, U., Mack, L., Yun, J., & Kim, S. H. (2011). Predicting the timing of cherry blossoms in Washington, DC, and mid-Atlantic states in response to climate change. *PLoS ONE 6*, e27439.

Clow, D. L. (2010). Changes in the timing of snowmelt and streamflow in Colorado: A response to recent warming. *Journal of Climate, 23,* 2293–2306. doi:10.1175/2009JCL12951.1

Cox, R. (2009). *Environmental communication and the public sphere* (2nd ed.). Thousand Oaks, CA: Sage.

Delach, A., & Matson, N. (2010). *Climate change and federal land management.* Washington, DC: Defenders of Wildlife.

Elsner, M. M., Cuo, L., Voisin, N., Deems, J. S., Hamlet, A. F., Vano, J. A., . . . Lettenmaier, D. P. (2009). Global warming: Implications for the recovery of the greenback cutthroat trout in Rocky Mountain streams. In S. J. Cooney (Ed.), *Implications of 21st century climate change for the hydrology of Washington State* (pp. 1–54). Fort Collins: Colorado State University.

Falk, J. H. (2005). Free-choice environmental learning: Framing the discussion. *Environmental Education Research, 11,* 265–280.

Falk, J. H., & Dierking, L. D. (2002). *Lessons without limits: How free choice learning is transforming education.* Walnut Creek, CA: AltaMira Press.

Falk, J. H., & Heimlich, J. E. (2009). Who is the free-choice environmental learner? In J. H. Falk, J. E. Heimlich, & S. Foutz (Eds.), *Free-choice learning and the environment* (pp. 23–37). Lanham, MD: AltaMira Press.

Fullilove, M. T. (1996). Psychiatric implications of displacement: Contributions from the psychology of place. *American Journal of Psychiatry, 153*(12), 1516–1523.

Gonzalez, P., Neilson, R. P., Lenihan, J. M., & Drapek, R. J. (2010). Global patterns in the vulnerability of ecosystems to vegetation shifts due to climate change. *Global Ecology and Biogeography, 19*(6), 755–768. doi:10.1111/j.1466-8238.2010.00558.x

Graber, D. A. (1976). *Verbal behavior and politics.* Urbana: University of Illinois Press.

Hanel, J. (2009, August 25). Sens. McCain, Udall sound alarm on warming national parks. *The Durango Herald.* Retrieved from https://durangoherald.com/articles/5588

Hess, J., Malilay, J., & Parkinson, A. J. (2008). Climate change: The importance of place and places of special risk. *American Journal of Preventative Medicine, 35*(5), 468–478.

Hicke, J. A., Logan, J. A., Powell, J., & Ojima, D. S. (2006). Changing temperatures influence suitability for modeled mountain pine beetle (Dendroctonus ponderosae) outbreaks in the western United States. *Journal of Geophysical Research, 111,* 1–12. doi:10.1029/2005JG000101

Huppert, D. D., Moore, A., & Dyson, K. (2009). Impacts of climate change on the coasts of Washington State. In *The Washington climate change impacts assessment: Evaluating Washington's future in a changing climate* (pp. 285–309). Seattle: Climate Impacts Group, University of Washington. Retrieved from http://cses.washington.edu/db/pdf/wacciach8coasts651.pdf

Huse, S. (2012). The retreat of Exit Glacier. Retrieved from https://www.nps.gov/kefj/learn/nature/upload/The%20Retreat%20of%20Exit%20Glacier.pdf

Klein, E., Berg, E.E., & Dial, R. (2005). Wetland drying and succession across the Kenai Peninsula Lowlands, south-central Alaska. *Canadian Journal for Restoration, 35,* 1931–1941. doi:10.1139/X05-129

Kola-Olusanya, A. (2005). Free-choice environmental education: Understanding where children learn outside of school. *Environmental Education Research, 11*(3), 297–307.

Kyle, R.E., & Brabets, T.P. (2001). *Water temperature of streams in the Cook Inlet Basin, Alaska, and implications of climate change* (Report No. 01-4109). Anchorage, AK: U.S. Geological Survey, U.S. Department of the Interior. Retrieved from https://pubs.usgs.gov/wri/2001/4109/wri01-4109.pdf

La Sorte, F.A., & Thompson, F.R. (2007). Poleward shifts in winter ranges of North American birds. *Ecology, 88*(7), 1803–1812.

Lee, T.M., Markowitz, E.M., Howe, P.D., Ko, C., & Leiserowitz, A.A. (2015). Predictors of public climate change awareness and risk perception around the world. *Nature Climate Change, 5,* 1014–1020.

Leiserowitz, A., Maibach, E., & Roser-Renouf, C. (2009). *Global warming's six Americas.* New Haven, CT: Yale Project on Climate Change Communication.

Leiserowitz, A., Maibach, E., Roser-Renouf, C., & Smith, N. (2011) *Global warming's six Americas, May 2011.* New Haven, CT: Yale Project on Climate Change Communication.

Leiserowitz, A., Smith, N., & Marlon, J.R. (2010). *Americans' knowledge of climate change.* New Haven, CT: Yale Project on Climate Change Communication. http://environment.yale.edu/climate/files/ClimateChangeKnowledge2010.pdf

Lorenzoni, I., & Pidgeon, N.F. (2006). Public views on climate change: European and USA perspectives. *Climatic Change, 77*(1–2), 73–95.

Maibach, E., Wilson, K., & Witte, J. (2010). *A national survey of television meteorologists about climate change: Preliminary findings.* Fairfax, VA: George Mason University, Center for Climate Change Communication.

McMahon, S.M., Parker, G.G., & Miller, D.R. (2010). Evidence for a recent increase in forest growth. *Proceedings of the National Academy of Sciences, 107*(8), 3611–3615. doi:10.1073/pnas.0912376107

Millar, C.I., Westfall, R.D., & Delany, D.L. (2004). Response of subalpine conifers in the Sierra Nevada, California, USA, to 20th-century warming and decadal climate variability. *Arctic, Antarctic, and Alpine Research, 36*(2), 181–200.

Moritz, C., Patton, J.L., Conroy, C.J., Parra, J.L., White, G.C., & Beissinger, S.R. (2008). Impact of a century of climate change on small-mammal communities in Yosemite National Park, USA. *Science, 322*(5899), 261–264. doi:10.1126/science.1163428

Moser, S.C., & Dilling, L. (2007). Toward the social tipping point: Creating a climate for change. In S.C. Moser & L. Dilling (Eds.), *Creating a climate for change: Communicating climate change and facilitating social change* (pp. 491–516). New York, NY: Cambridge University Press.

Mote, P.W. (2003). Trends in snow water equivalent in the Pacific Northwest and their climatic causes. *Geophysical Research Letters, 30*(1601), 1–4. doi:10.1029/2003GL017258

Mote, P.W., Hamlet, A.F., Clark, M.P., & Lettenmaier, D.P. (2005). Declining mountain snowpack in western North America. *Bulletin of the American Meteorological Society, 86*(1) 39–49. doi:10.1175/BAMS-86-1-39

Mote, P.W., & Salathé, E.P., Jr. (2009). Future climate in the Pacific Northwest. *Climatic Change, 102*(1), 29–50.

Mullen, K., Newman, G., & Thompson, J.L. (2014). Facilitating the development and evaluation of a citizen science website: A case study of repeat photography and climate change in southwest Alaska's national parks. *Applied Environmental Education and Communication, 12*(4), 261–271.

O'Neill, S., & Nicholson-Cole, S. (2009). "Fear won't do it": Promoting positive engagement with climate change through visual and iconic representations. *Science Communication, 30*(3), 355–379.

Orr, D.W. (1994). *Earth in mind: On education, environment, and the human prospect.* Washington, DC: Island Press.

Raymond, C.M., & Brown, G. (2011). Assessing spatial associations between perceptions of landscape values and climate change risk for use in climate change planning. *Climatic Change, 104,* 653–678. doi:10.1007/s10584-010-9806-9

Reynolds, T.W., Bostrom, A., Read, D., & Morgan, M.G. (2010). Now what do people know about global climate change? Survey studies of educated laypeople. *Risk Analysis, 30*(10), 1520–1538. doi:10.1111/j.1539-6924.2010.01448.x

Salazar-Halfmoon, V. (2010). *Vanishing treasures 2010 year-end report: A climate of change.* Denver, CO: National Park Service.

Scavia, D., Field, J.C., Boesch, D.F., Buddemeier, R.W., Burkett, V., Cayan, D.R., & Titus, J.G. (2002). Climate change impacts on U.S. coastal and marine ecosystems. *Estuaries, 25*(2), 149–164.

Schweizer, S., Davis, S.K., & Thompson, J.L. (2013). Changing the conversation about climate change: A theoretical framework for place-based climate change engagement. *Environmental Communication: A Journal of Nature and Culture 7*(1), 42–62.

Sobel, D. (2004). *Place-based education: Connecting classrooms and communities.* Great Barrington, MA: Orion Society.

Stedman, R.C. (2003). Sense of place and forest science: Toward a program of quantitative research. *Forest Science, 49*(6), 822–829.

Stern, N.H. (2007). *The economics of climate change: The Stern review.* Cambridge, UK: Cambridge University Press.

Thomashow, M. (2002). *Bringing the biosphere home: Learning to perceive global environmental change.* Cambridge, MA: MIT Press.

Thompson, J., Davis, S., & Mullen, K. (2013). Climate change communication campaign planning: Using audience research to inform design. *The George Wright Forum, 30*(2), 182–189.

van Mantgem, P.J., Stephenson, N.L., Byrne, J.C., Daniels, L.D., Franklin, J.F., Fulé, P.Z., ... Veblen, T.T. (2009). Widespread increase of tree mortality rates in the western United States. *Science, 323*(5913), 521–524. doi:10.1126/science.1165000

Wanless, H. R., & Vlaswinkel, B. M. (2005). *Coastal landscape and channel evolution affecting critical habitats at Cape Sable, Everglades National Park.* Miami, FL: University of Miami.

Woodhouse, J. L., & Knapp, C. E. (2000). *Place-based curriculum and instruction: Outdoor and environmental education approaches.* Retrieved from ERIC database. (ED448012)

A Wildlife Expeditions participant uses a spotting scope to get an up-close look at wildlife from a safe and ethical viewing distance. Courtesy of Teton Science Schools.

6

Place-Based Education at Teton Science Schools

Inspiring Curiosity, Engagement, and Leadership in National Parks and Beyond

KEVIN KRASNOW, NATE MCCLENNEN, AMANDA KERN, PATRICK LEARY, AND GREG PECK

INTRODUCTION

Place-Based Education

Place-based education, perhaps a new term, is not a new practice. For as long as humans have learned from one another, education has been place-based—where "curriculum" (what is taught) is directly connected to the local community. Originally, this was out of necessity, as survival depended on understanding the local flora and fauna, cultural norms, landscapes, and seasonal changes of a particular area. Over the past 125 years or so, formal education has largely separated from place with standardized curriculum for all students that often ignored local economy, culture, and ecology. However, at this same time, "informal" educators frequently immersed participants in local places to provide an experiential model that founded learning on the richness of the surrounding community.

Teton Science Schools (TSS) was established in 1967 as a program for students in the vast outdoor classroom of Jackson Hole, Wyoming. Very quickly, Grand Teton National Park became a key partner, with a permanent campus located within the park boundaries. Over the past twenty-five years, TSS has embraced place-based education and defines the practice as *connecting learning and community to increase student engagement, educational outcomes, and community impact*. We define *place* broadly to encompass *ecological, cultural, and economic* components of a community. In 2016 TSS adopted a new mission statement: *"Inspiring curiosity, engagement, and leadership through transformative place-based education."* This mission change allowed TSS to better encompass the work of a greatly expanded organization, which now includes two independent schools, a graduate program, and a teacher professional-development center in addition to its original field education programs.

As the concepts of *deeper learning*, *project-based learning*, *personalized learning*, and *competency-based learning* push to the forefront in a post–No Child Left Behind era, organizations such as national parks are beginning to play a bigger role as providers of learner experience within "formal" education spaces. This blending of informal and formal is a unique opportunity for the National Park Service (NPS) to extend its educational mission into spaces typically not served by Department of Interior agencies.

TSS Place-Based Educational Philosophy

The educational value of place has always been central to TSS, which is situated in a scenic and ecologically active corner of Grand Teton National Park. Over our fifty-year history, we have refined our educational approach and now seek to share what we have learned about how to authentically engage one's place and the advantages of doing so. We have developed six principles to guide place-based education design for practitioners: (a) local to global context, (b) learner centered, (c) inquiry based, (d) design thinking, (e) community as classroom, and (f) interdisciplinary approach (see www.tetonscience.org/about/place-based-education/).

In the time we have been implementing placed-based education, we have seen positive results from our approach. Students acquire an improved appreciation and understanding for this place and themselves, and gain new tools to better understand, engage, and impact their own community. We believe that effective place-based education increases student and teacher engagement, boosts academic outcomes, and results in meaningful community impact.

In the following case study examples, we illustrate some impacts of our diverse place-based programs in national parks, detail our recent educational research, and share ideas and best practices to further understanding of place-based education. We recognize that our examples come from a region of immense natural beauty and complexity, but we also strongly believe effective placed-based education can happen anywhere, anytime. Every place has a unique history and story, and we believe engaging students in their own community has the power to transform education.

CASE STUDIES
Teton County Fifth Graders and Kelly Warm Springs Invasive
Fish Study: Increasing Student Engagement and Community Impact

Kelly Warm Springs is an easily accessible, front-country thermal feature in Grand Teton National Park. The spring runs into Ditch Creek and ultimately the Snake River, which is habitat to the greater ecosystem's only native trout—the Yellowstone cutthroat trout. While there is a long history of illegal dumping of warmwater aquarium fish, the urgency of this problem increased when the Grand Teton National Park scientists discovered invasive species like goldfish, madtoms, and bullfrogs in Ditch Creek, only ten meters from the Snake River. The nonnative species prey on native fish eggs and juveniles, deplete food sources, and spread disease.

At the start of the work with Teton County fifth-grade students, the scientists needed more information on species inventory and presence of diseases. This citizen science project came about as part of a multi-pronged project that included interagency research on invertebrates and bullfrogs, an inventory of fish species present, and public outreach. There were four sampling locations, two in the spring source and two downstream in the outflow. Students took turns being a "fish biologist" who identified species, a "fish doctor" who identified diseases, data recorder, photographer, and fish handler.

The students collected data regarding the following:

- fish species and length
- native/nonnative status
- aquarium diseases present; for example, eye bulge, dropsy, fin rot, or ulcers
- air temperature
- water temperature and chemistry
- aquatic vegetation

Data collection was designed to capture baseline information about the fish in the spring. In addition, instructors facilitated observations around the warm spring and then students chose to explore questions such as the following:

- Are there more native fish in the spring source or in the outflow?
- Do native and nonnative fish both show signs of the aquarium diseases?

For their closing project, students presented their data, compared it with other groups, and drew conclusions. The students' engagement in this project was evident: one student group working in the summer of 2015 were so engaged they didn't notice an approaching thunderstorm. "This school is different than [mine]," said a ten-year-old. "At my school [we] always stay inside, except for recess. Here, we are outside and learning."

The Kelly Warm Springs project has had powerful results both for students' science skills and public outreach. When polled, nearly every student group "[knew] someone who has dumped a fish in Kelly Warm Springs." Throughout the course of analyzing data, students drew powerful conclusions. For example, one student, whose father makes a living as a local angler and outfitter, recognized that invasive fish could have a direct impact on his family's livelihood. When rangers visited the classroom weeks later, students could still explain their experiments. Now that the program is several years old, rangers have encountered younger siblings who know about the dangers of dumping fish in Kelly Warm Springs.

Journeys School Group 4 Project: Increasing Student Knowledge of and Connection to Place

TSS's Journeys School is an International Baccalaureate (IB) world school serving students pre-K to grade twelve. While the school's eleventh- and twelfth-grade science curriculum is explicitly dictated, an exception is the Diploma Program's Group 4 Project. The Group 4 Project is broadly focused on students developing an understanding of the relationship between scientific disciplines and scientific methods. Each February, students complete the project by spending four days and three nights in Grand Teton National Park conducting original scientific research.

The students develop research questions and hypotheses (with teacher guidance) that reflect their curiosities about their place. In small groups, students work together to establish research methods for collecting quantitative data to support or refute their hypotheses. Throughout the project, students are given complete autonomy, knowing that they will present their findings to a group of peers and faculty who will critique their work. Off campus and away from traditional pressures of school, extracurriculars, and family, students' natural curiosity flourishes. Examples of past project titles include (a) *The Effects of Snow Depth on Solar Panel Efficiency*; (b) *How Emissions from an Idling Car Affect Pollutant Concentrations in Snow*; and (c) *Social Pressure versus Stress in High School Students.*

Student feedback on Group 4 Project experiences is consistent in their enjoyment of working with peers and having the freedom to choose their own topics. A student noted that "the freedom to choose an experiment in [any discipline] combined with the . . . availability of the natural world made this project an unforgettable learning experience." In a post–high school reflection, a 2017 Journey School graduate endorsed this idea further, saying "The Group 4 Project connects students, allowing them to plan experiments that perfectly involve science, the natural world, and students together in one." In this way, Grand Teton National Park and the TSS Kelly Campus provide students the ideal environment to ask and answer questions about their surroundings and connect more deeply with their place.

Educational Research with Columbia School District:
Exploring Students' Attitudes and Understanding of Science

As an initial foray into emerging educational research we wanted to study the impact of our long-standing field education programs for visiting middle and high school students. We chose to examine the effects of a six-day program that includes visits to both Grand Teton and Yellowstone National Parks on students from the Columbia (Missouri) School District during the summer of 2016. This district has been coming to TSS since 2012 and was also interested in this research. We partnered with researchers from the University of Wyoming and Utah State University and sought to assess student growth in three areas: student attitudes about science (Moore & Foy, 1997), student self-efficacy (Faber et al., 2013), and students' understanding of the Nature of Science (Next Generation Science Standards Lead States, 2013, appendix H).

We employed a pre- and postsurvey design in which we developed a survey instrument to measure each outcome of interest. Each participating student completed the presurvey shortly before their visit to TSS and then completed the postsurvey on their final morning at TSS. When we compared students' presurvey responses with their postsurvey responses, we found significant growth on all of the following assessment items ($n = 206$):

Attitudes about Science

- I am sure of myself when I do science.
- I know I can do well in science.

Self-Efficacy

- I am confident I can lead others to accomplish a goal.
- I am sure I can help my peers.
- I feel comfortable including others' perspectives when making decisions.
- I am confident I can make changes when things do not go as planned.

Understanding of the Nature of Science

- Scientific knowledge is based on empirical evidence.
- Scientific knowledge is open to revision in light of new evidence.

These findings confirmed anecdotal observations and feedback we have received over time and echoed written feedback received from this group's teachers and chaperones. They also indicate trends that match the literature and indicate that our programs have the potential to teach academic content, impact students' attitudes about science, increase their effectiveness in working with and leading others, and increase students' understanding of the Nature of Science (Houseal, Abd-El-Khalick, & Destefano, 2014). We aim to continue this research, but we also seek to understand better how various components of our programs impact student outcomes and look forward to rigorously testing our hypotheses about place-based education and sharing our methods and findings.

Wildlife Expeditions: Inspiring Curiosity through Ecotourism

Wildlife Expeditions is the only nonprofit wildlife tour provider in Jackson, Wyoming. In 1999, Wildlife Expeditions joined TSS and presently we connect approximately thirty-five hundred visitors to Grand Teton National Park and Yellowstone National Park annually. Much like other program areas, the purpose of Wildlife Expeditions is to inspire curiosity, help people become more inquisitive, and provide experiences in national parks that allow people to gain a better understanding of interconnections between local and global communities. We accomplish this through half-, full-, and multiday wildlife tours throughout the year.

During tours, the guides are charged with providing memorable, educational experiences for national park visitors. We have found that the most impactful way to achieve this is by using place-based education practices. For example, when viewing pronghorn antelope feeding in sagebrush ecosystems, guides set up spotting scopes and allow guests to observe and verbalize what they see. Next, the guide has guests observe their surroundings to find the plants the pronghorn are eating. They are encouraged to observe the plants up close, smell them, and imagine eating only grasses and sagebrush—the latter being a plant that few animals are capable of eating. At this point, a guide may begin teaching about Wyoming's sagebrush ecosystems, which tend to be where prime oil and gas development opportunities exist. In this way, the lesson merges natural history with culture and economics. This example highlights how an informal conversation built on initial observations of pronghorn and sagebrush and can be extended into complex relationships among ecological, cultural, and economic forces concerning pronghorn habitat, sagebrush ecosystem ecology, and oil and gas development. Discussions like these are designed to relate factual information and prompt guests to consider different perspectives.

In addition to facilitating dialogue, some tours involve hands-on experiences. A favorite Wildlife Expeditions activity is to make plaster casts of animal tracks discovered during a tour. To find well-imprinted tracks, guests must pay close attention to their surroundings. Through the process of searching, guests practice observation skills and establish a connection with the ecosystem. Afterward, guests are invited to keep their plaster cast as a reminder of connections they made during their national park visit.

CONCLUSION

The Future of Place-Based Education: National Parks and Beyond

These case studies highlight the six design principles of place-based education. Participants understand *local-to-global* context by examining local invasive species or transferring Nature of Science skills from one region to another. Place-based experiences are *learner centered* when high school students design winter ecology projects or elementary school students write research questions. *Inquiry-based* and *design-thinking* approaches are explicitly used, from visitors learning more about the dynamic Greater Yellowstone Ecosystem to students proposing solutions for invasive fish challenges in Kelly Warm Springs. By using the *community as a classroom*, students see the Jackson Hole community as an exemplar for any community—where the intersection of economy, culture, and ecology plays a critical role in both understanding and solution making. Finally, students at TSS experience the *interdisciplinary approach* of place-based education through interdisciplinary projects as students explore how a culture of fire suppression impacts ecosystems. These experiences build a worldview that each place is understood and improved through a number of interdisciplinary lenses.

US education is at a crossroads where "informal" learning organizations are beginning to have a role in formal K–12 spaces. The NPS and many other nonprofit and agency organizations have long been relegated to school visits and summer, after-school, and service learning programs rather than embedded in educational partnerships. At TSS, we foresee a future where *place-based education* could be core to every school. With countless informal education partners, including those associated with the NPS, students may begin to see that their classroom is not just a space in the traditional school, but the entire community and its economic, ecological, and cultural components.

Using technological advances and strong evidence of personalized and project-based learning, schools are realizing that "anytime, anywhere" learning is the next step beyond the "one size fits all" industrial model. As schools link learning outcomes with experiences outside the classroom, organizations and agencies like TSS and the NPS are poised to play a much larger role in formal education spaces.

Acknowledgments

The authors are grateful to Megan Kohli, Program Manager: Youth, Community Engagement & Volunteers at Grand Teton National Park, for her contribution to this chapter and place-based education efforts in the Tetons.

References

Faber, M., Unfried, A., Wiebe, E. N., Corn, J., Townsend, L. W., & Collins, T. L. (2013, June). *Student attitudes toward STEM: The development of upper elementary school and middle/high school student surveys.* Paper presented at the 120th American Society of Engineering Education Conference, Atlanta, GA.

Houseal, A. K., Abd-El-Khalick, F., & Destefano, L. (2014). Impact of a student-teacher-scientist partnership on students' and teachers' content knowledge, attitudes toward science, and pedagogical practices. *Journal of Research in Science Teaching, 51*(1), 84–115.

Moore, R. W., & Foy, R. L. H. (1997). The scientific attitude inventory: A revision (SAI II). *Journal of Research in Science Teaching, 34*(4), 327–336.

Next Generation Science Standards Lead States. (2013). *Next Generation Science Standards: For states, by states.* Washington, DC: National Academies Press.

Hiking in the snow. Courtesy of Ana Houseal.

7

Three-Dimensional Learning

"Upping the Game" in Citizen Science Projects

ANA K. HOUSEAL

INTRODUCTION

Citizen science projects continue to increase in popularity. National parks, with their abundance of wilderness, accessibility for the general public, and recent focus on education, are a natural place for these types of projects—and many are being conducted. However, while many citizen science projects are participatory in nature, the type of participation is generally limited to data collection of specific protocols and involves citizen scientists in very limited ways. A Framework for K–12 Science Education (National Research Council [NRC], 2012), developed to guide science standards for formal educational settings, proposes that learning science is dependent on three dimensions: science content, scientific processes, and overarching ideas that span the discipline. When these three are integrated, the scientific enterprise can be understood in a more realistic and authentic manner. This chapter proposes the use of this framework to increase the richness of participant involvement in citizen science projects, especially in nonformal settings.

"I can't leave yet, I am not done collecting our data."
(Sixth-grade student)

It's approximately −7°C (20°F) and snowing as the fifth- and sixth-grade students from Denver are schlepping scientific data collection tools on their backs, snowshoeing through deep drifts heading to a remote hot spring, guided by an Expedition: Yellowstone! National Park Service (NPS) ranger, some university researchers, and their teachers. After they arrive and collect data (from atmospheric conditions to qualitative descriptions) for an ongoing geomicrobiology study, they dive into answering their own research questions. These questions were developed during several classroom-based sessions, which took place prior to and after arriving at Yellowstone National Park. Groups of students explored questions focused on phenomena such as the effects of hot spring water temperature on the color of microbial communities; describing conditions that seem to be conducive to the presence of "calcite ice"; and the effects of hot spring flow rate on pH. Students persevered through nasty winter conditions until all their data had been collected, and felt very accomplished.

After returning to the lodge, students changed clothes, began analyzing data, and developed evidence-based explanations. In spite of the prior preparation and fieldwork, it was this last step of communicating their findings that was the most critical in their understanding of the scientific process. As they shared their results, it became apparent that their understanding of the hot springs system was cemented during their presentations.

The students participating in the residential environmental educational experience, Expedition: Yellowstone!, were conducting citizen science research. As a growing field with an abundance of different types of projects, citizen science, when developed and executed thoughtfully, can serve science learning in numerous ways. However, some of the methods used within these projects limit how much science is actually being done by participants and thus limit the richness of their scientific experiences. This Expedition: Yellowstone! vignette will be used as an example of a rich citizen science project.

CITIZEN SCIENCE

Citizen science holds several names in the literature. In the late 1990s and early 2000s, Student-Scientist-Partnerships and Student-Teacher-Scientist-Partnerships covered many formal educational partnerships. In the nonformal realm, Public Participation in Science Research is a term that has more recently become popular. Definitions of citizen science contain some basic similarities. Key elements are that the projects are

- partnerships between scientists and nonscientists,
- used to answer science research questions,
- designed to include educational goals for the nonscientists, and
- projects that tend to fall along a continuum of models depending on the scientific engagement of participants (Becker-Klein, Peterman, & Stylinski, 2016; Bhattacharjee, 2005; Bonney et al., 2009; Jordan, Ballard, & Phillips, 2012; Shirk et al., 2012; Silvertown, 2009; Vitone et al., 2016).

Silvertown (2009) argues that citizen science projects have been around since science was a side endeavor of professionals in other areas or those who could afford to do it (e.g., Benjamin Franklin and Charles Darwin). In this way, science has relied on amateurs' observation skills for a long time, especially in areas such as archeology and natural history. Silvertown posited that more recently the popularity of these projects is growing due to three critical factors. First, the availability of social media and electronic technologies ease constraints that used to limit communication about, and collecting and submitting data for, projects. Second, in conjunction with the decrease in research funding, the general public can be leveraged as a source of both free labor and much needed financial support. Finally, a focus on broader impact requirements of national funding agencies, which require better connections between the research and the general public, has sparked development of these types of projects. The idea that appreciation of science could be enhanced by participation is another factor cited.

Shirk et al. (2012) proposed that citizen science projects could be categorized into five models. Two of them focus on contracted scientist-community collaboration and independent noncredentialed researcher contributions. The other three, the *contributory, collaborative*, and *cocreated* models, can be used to describe projects that involve the general public and scientific researchers and teams. More specifically, contributory citizen science projects primarily limit the involvement of volunteers to the collection of data (usually through observation, identification, and monitoring). Collaborative citizen science projects expand on data collection to include developing explanations, designing data collection methods, and analyzing data. Cocreated projects include the inquiry skills from the other two models. They also provide opportunities for the public to define the research question, gather information to support the study rationale, interpret data to make conclusions, disseminate results, and post questions for future study (Becker-Klein et al., 2016).

Currently, more citizen science projects are contributory in nature, though many more collaborative and cocreated projects are being developed (Becker-Klein et al., 2016). The initial example in this chapter would be defined as a cocreated project, as students and teachers were involved in the decision-making and scientific processes to answer questions they had developed.

A Framework for K–12 Science Education and Three-Dimensional Learning

While science and engineering are pervasive in modern life, many people lack basic knowledge in these fields (NRC, 2012). A Framework for K–12 Science Education (subsequently referred to as *the Framework*) sought to address this deficiency. This framework includes three main dimensions critical to developing scientific understanding, often referred to more broadly as scientific literacy. The disciplinary core ideas (science content; DCIs), crosscutting concepts (big ideas that connect disciplines; CCCs), and scientific and engineering practices (processes for conducting science; SEPs) work together in this model to build understanding and connect the often-disparate aspects of science (NRC, 2012).

Curriculum and strategies that are being developed that explicitly work on merging the dimensions have been dubbed *three-dimensional learning* (3D learning). In addition to changing K–16 science education, these dimensions and their implications match better the type of learning (in any area) that takes place outside of formal educational settings. Thus, it can be a good model to examine in terms of citizen science projects.

The Framework was conceived to push forward a vision for science education "in which students, over multiple years of school, actively engage in scientific and engineering practices and apply crosscutting concepts to deepen their understanding of the core ideas in these fields" (NRC, 2012, p. 10). Figure 7.1 illustrates this model of scientific inquiry, which explicitly combines the three dimensions. While the purpose of the Framework was to translate research for the development of new educational science standards, I propose that this model demonstrates how overlapping the dimensions in different ways can be useful in thinking about making citizen science projects more robust and rich for participants, as demonstrated by the Expedition: Yellowstone! example.

Integration of
CONTENT, BIG IDEAS, AND PROCESS

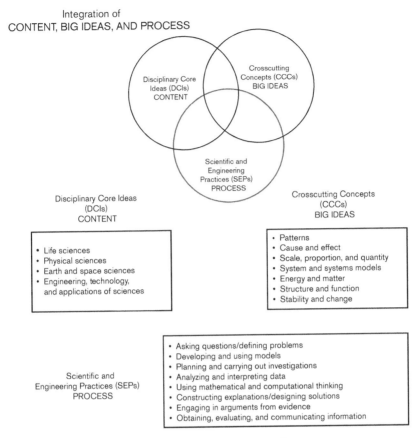

FIGURE 7.1. Model of the three dimensions of science learning.

The overlapping areas in this model are the most critical. While science involves all three of these areas (and a fourth not represented, the Nature of Science [NOS], which defines how science is different from other human endeavors), often "school science" and other learning opportunities teach them separately, or not at all. For example, reading about the final results of scientific inquiry or learning about theories and laws through text (DCIs) leaves out the idea that there are varied processes used to get to this understanding (SEPs) and big ideas that both inform and connect these across the disciplines (CCCs). The most powerful spot in this model is in the center, where participants can be exposed to science in a way in which it is explicit that they cannot only be consumers of scientific knowledge, by developing understanding about the content and the big ideas, but also producers of this knowledge, by participating in the practices to gain this understanding. Even though the center represents three-dimensional learning and is a desired end point, a good starting point could be in the two-dimensional overlapping areas. As an example, asking scientific questions and creating evidence-based claims about patterns in previously gathered data could engage learners in a deeper understanding of science.

Another key idea posited by the Framework is that participation in the practices of science is the best way to understand science. The research document that led to the development of the Framework, *Taking Science to School* (NRC, 2007), stated, "To understand science, one must use science and do so in a manner that reflects the values of scientific practice" (p. 40). In this way, a key aspect of this model and the Framework's vision is the idea of *who* is *doing* the work within each of the dimensions. In other words, to gain scientific understanding, it is not enough to be provided with protocols to gather data or a formula or format for analysis. To develop understanding, participants need to be involved in making decisions about what and how data are to be collected and analyzed. This leads to another key consideration: balancing the needs of scientists to have quality data with the needs of students to learn how to conduct science. The following section will outline how citizen science does—and could further—address each of the dimensions in this framework.

Scientific and Engineering Practices

Contributory projects, as defined by Shirk et al. (2012), often fall just outside of the SEP circle of this model because scientists often offer specific questions and protocols. While standardized protocols are a staple of scientific investigation, the benefit of this system skews toward the scientist, who is gaining data. If students are to meaningfully engage in the practices of science (the SEP circle), a tension lies in balancing the needs of the scientists to have quality data and the desire to have the experience provide for participants understanding of science and the processes therein. In addition, there are many practices outside of the traditional *planning and carrying out investigations* practice, in which data collection lies, that can be used to engage participants. For example, the SEP *engaging in arguments from evidence* can be used as a vehicle to engage participants with the data collected and claims that can and cannot be made.

Disciplinary Core Ideas

Many citizen science projects provide background in the content area for participants. In nonformal settings, participants may decide to engage in specific citizen science projects based on their prior interest in a particular content area. For example, amateur bird watchers may engage in a citizen science project that focuses on identification or population count because of a previous interest. This type of contributory citizen science project may provide additional content background that would fall within the DCIs; however, the onus is on the participant to be motivated beyond the data gathering and reporting to dig deeper. In many cases, the scientists have a depth of content knowledge that informs their decision-making within the project. Making that accessible to participants could help enhance understanding of the project and science as a knowledge-building endeavor.

Crosscutting Concepts

It is a relatively new idea to use CCCs as overarching connections among the science disciplines and they are rarely explicitly present in citizen science projects. CCCs can be used to strengthen connections among aspects within and among citizen science projects, and as the big ideas that span disciplines and contexts, they can be used to help with understanding connections that might otherwise be difficult to access. For example, CCCs *systems and systems models* and *energy and matter* can be used to understand the flow of energy within a physical system, such as the conversion of energy in a Rube Goldberg machine, and likewise within a trophic food chain. Energy in both of these systems, as it is converted, becomes less usable in each successive link. Likewise, studying *patterns* in populations of migrating butterflies might lead to greater understanding of this species and other migrating species. While citizen science projects will vary in terms of the CCCs they access, this dimension offers a broader perspective.

Dimension Complexity

As citizen science project designers and implementers consider the three dimensions of learning, it's important that they note the practices and crosscutting concepts are more complex than they seem. For example, SEP #1, *asking questions*, is often misunderstood by teachers to cover the asking of any question during teaching. In fact, as mentioned before, *who* is doing the asking is of utmost importance, and the type of questions being asked (in this case, scientific research questions—or questions that help define scientific ideas) is also critical. *Patterns* (CCC #1) has a continuum of complexity ranging from a simple repeating pattern to many varied, complex patterns. In addition, the idea that exploring patterns in one system or context can help explain another system is an example of the way CCCs aid with scientific thinking and understanding. Moreover, *developing and using models* (SEP #2) and *systems and systems models* (CCC #4) are examples of interrelated dimensions. A sophisticated understanding of models and modeling (the types and uses of various models, from conceptual to visual to mathematical) takes time and practice to develop. Finally, the DCIs that must be engaged to provide the context of the project also span a continuum, from surface-level subject matter to deeply complex disciplinary content. Depending on the model of the project, there will be a range to the depth of knowledge needed by the participants and an expansion in the idea of who will seek the information. Understanding the dynamics of these dimensions will be critical to engaging scientists and nonscientists effectively.

With the complexity of the dimensions in mind, and an understanding that they are not intended to be accessed all at once, we should think carefully about which ones lend themselves best to citizen science projects. Some possibilities lie in table 7.1.

Returning to the Exploration: Yellowstone! example, the Denver students participated in a project that in many ways exemplified this shift. The Student-Teacher-Scientist-Partnership consisted of three separate parts. In part I, a photo-point collection, students took photos in specific locations and added to a database of photos taken within Mammoth Hot Springs that recorded the changes over time of specific spots. In this case, they were making observations of the system, gathering data, and using that evidence to back up claims about travertine growth.

In a second aspect of the project, part II, students learned to use specific tools (infrared surface temperature thermometers, pH probes, a geologic model of hot springs systems, and data collection grids, to name a few) to collect a specific set of data for an ongoing geomicrobiological study of Mammoth Hot Springs. Finally, in part III, using some of the same tools they learned to use for the scientists' data collection, students developed their own scientific questions, then collected, analyzed, and developed explanations for them—using scale and systems

TABLE 7.1 MODELS OF CITIZEN SCIENCE PROJECTS AND POSSIBLE CONNECTIONS TO 3D LEARNING

Citizen science models/dimensions	Disciplinary core ideas (DCIs)—Content	Scientific and engineering practices (SEPs)—Process	Crosscutting concepts (CCCs)—Big ideas
Contributory	Content that matches the citizen science project; a variety of sources may be provided by scientists, mostly for consumption and developing a depth of understanding	• Planning and carrying out investigations (e.g., making observations; connecting to scientific questions and including ways for participants to evaluate observations) • Analyzing and interpreting data (e.g., collecting and reporting data; connecting explicitly to CCCs and DCIs)	Make connections related to scientific questions using • patterns • cause and effect
Collaborative	Content provided by both scientists and participants; focused on encouraging participants to make connections, seek new sources, and develop new ideas	• Planning and carrying out investigations (e.g., used to develop a deeper understanding of data collection and analysis) • Developing and using models • Engaging in arguments from evidence	Use these CCCs following decisions about data collection and/or analysis: • patterns • cause and effect • systems and systems models
Cocreated	Increasing depth; involvement in the consumption and dissemination of new scientific understanding	Could use any of the following (or all SEPs): • planning and carrying out investigations • developing and using models • obtaining, evaluating, and communicating information	Use and apply the following: • patterns • cause and effect • systems and systems models Make explicit connections among all three dimensions

NOTE: This is not intended to be an exhaustive list, but instead a starting point.

models as a way to understand their and their classmates' results. The data collection for scientists portion (part II) could be defined as *contributory*, the photo-point portion of this project (part I) could be defined as *collaborative*, and the final project (part III) could be defined as *cocreated*.

Within the 3D model, part I would fall into the crossover between DCIs and SEPs, using *earth and space sciences* and *analyzing and interpreting data*. Since part II was about gathering specific data for someone else, it maps only within the SEP circle, focused on the data collection portion of *planning and carrying out investigations*. For part III, students were engaged at the center of the 3D model as they asked and answered their own questions, gathered and interpreted their data, and presented their findings. They specifically engaged with the DCIs *earth and space science* and/or *life science,* the SEPs *asking questions, planning and carrying out investigations, analyzing and interpreting data,* and *engaging in arguments from evidence*, and the CCCs *scale, proportion, and quantity* and *systems and systems models*.

RECOMMENDATIONS

Creating 3D learning opportunities, like the Expedition: Yellowstone! project, will take some thoughtful work. Some things to consider include evaluating current projects for ways to tweak the design such that aspects of the dimensions are explicitly present, as noted in table 7.1. When designing citizen science projects, pay careful attention to the following:

- who the participating audience is—youth, adults, ecosystem visitors, or locals;
- what type of project it is, and what kinds of questions are being asked;
- who the scientists are, what they are studying, and how the study can be connected to the learning audience before and after the project;
- how the dimensions support one another, and in which order they should appear for participants' understanding and contribution within the citizen science project;
- how the project will be framed by park staff, collaborating scientists, organizers, and participants;

- what type of citizen science model should be used—sometimes contributory or collaborative projects are appropriate, but do not try to force them to be cocreated; and
- how to challenge projects to be as multidimensional as possible.

It is possible to build multidimensional learning opportunities using these recommendations, and the Framework and 3D learning hold the potential to increase the richness of participant involvement in and learning about science within citizen science projects.

CITIZEN SCIENCE AND THE NPS

NPS units are logical places to situate citizen science projects and many partnerships are already in existence. Examples range from very large-scale operations—such as the National Geographic Bio Blitzes conducted in 120 NPS units—to smaller, more focused projects—like Monarch Butterfly Tagging in Smoky Mountain National Park (www .gsmit.org/CSMonarchTagging.html) and SeaWatch at Schoodic Point in Acadia National Park (www.schoodicinstitute.org/what-we-offer /public-opportunities/citizen-science-opportunities/seawatch-schoodic-point/). In addition, there is a stronger emphasis on the *broader impacts* for grant applications than in the past. In focusing on these broader impacts, funding agencies aim to engage the public in these scientific endeavors and "up the game" when it comes to sharing and disseminating scientific research. A logical choice for some of these grant proposals are citizen science projects. This is especially true for scientific questions related to or studying/needing large data sets (due to temporal or spatial scales), especially in the earth and life sciences.

CONCLUSION

Designing new and redesigning existing projects with the ideas presented in this chapter could have some exciting outcomes. For NPS visitors, citizen science projects could increase knowledge of and involvement in the scientific process, through the lens of a specific project. Potential outcomes for scientists could include both the gathering of data and deeper personal connections with participants through their mutual involvement in the dimensions. Finally, NPS units could realize potential outcomes in terms of greater visitor engagement and connections to funding opportunities by engaging in integrated citizen science partnerships.

References

Becker-Klein, R., Peterman, K., & Stylinski, C. (2016). Embedded assessment as an essential method for understanding public engagement in citizen science. *Citizen Science: Theory and Practice, 1*(1), 8. doi:http://doi.org/10.5334/citizen sciencetp.15

Bhattacharjee, Y. (2005). Citizen scientists supplement work of Cornell researchers. *Science, 308*(5727), 1402–1403. doi:10.1126/science.308.5727.1402

Bonney, R., Cooper, C.B., Dickinson, J., Kelling, S., Phillips, T., Rosenberg, K.V., & Shirk, J. (2009). Citizen science: A developing tool for expanding science knowledge and scientific literacy. *BioScience, 59*(11), 977–984.

Jordan, R.C., Ballard, L.H., & Phillips, T.B. (2012). Key issues and new approaches for evaluating citizen science learning outcomes. *Frontiers in Ecology and the Environment, 10*(6), 307–309. doi:10.1890/110280

National Research Council. (2007). *Taking science to school: Learning and teaching science in grades K–8.* Washington, DC: National Academies Press. doi:https://doi.org/10.17226/11625

———. (2012). *A framework for K–12 science education: Practices, crosscutting concepts, and core ideas.* Washington, DC: National Academies Press. doi:https://doi.org/10.17226/13165

Shirk, J.L., Ballard, H.L., Wilderman, C.C., Phillips, T., Wiggins, A., Jordan, R., ... Bonney, R. (2012). Public participation in scientific research: A framework for deliberate design. *Ecology and Society, 17*(2), 29. doi:http://dx.doi.org/10.5751/ES-04705-170229

Silvertown, J. (2009). A new dawn for citizen science. *Trends in Ecology & Evolution, 24*(9), 467–471. doi:https://doi.org/10.1016/j.tree.2009.03.017

Vitone, T., Stofer, K.A., Sedonia Steininger, M., Hulcr, J., Dunn, R., & Lucky, A. (2016). School of ants goes to college: Integrating citizen science into the general education classroom increases engagement with science. *Journal of Science Communication, 15*(01), A03.

Mountain Raingers pictured in front of the Laurel Gap backcountry shelter after having installed the rain gauge post at Big Cataloochee Mountain. Courtesy of Daniel Martin.

8

Mentoring Mountain Raingers

Beyond Basic Hydrological Field Research in the Great Smoky Mountains

DOUGLAS K. MILLER

INTRODUCTION

A high-elevation tipping bucket rain gauge network has been collecting rainfall data at remote locations in and near the Great Smoky Mountains National Park since 2007. Undergraduate field researchers at the University of North Carolina Asheville (UNCA) assisted in scouting sites, installing rain gauges, and maintaining and calibrating the gauges.

A typical day for the *Duke University–Great Smoky Mountains Rain Gauge Network* (Duke-RGN) team members was never very typical. Any number of factors, external (e.g., animals, insects, plants) and internal (e.g., drained battery, aging parts, insects) could prevent the accurate measuring and recording of rainfall accomplished through the counting of bucket tips. Bucket tips occur when the rain gauge bucket fills and *tips* to release collected rainwater. While the date and time of each tip are recorded with a battery-powered data logger housed inside the gauge funnel cover, the condition of the rain gauge and recording device must be verified periodically. There is an increased potential of gaps in data if there is a delay in visiting the rain gauge and recording device, especially if any problems arose since the previous visit. Thus, frequent rain gauge visits are necessary to ensure a continuous and high-quality data record. The team members hauled tools, monitoring and recording equipment, and personal gear to remote gauge locations in the Smoky Mountains. This team of researchers, hereafter called the

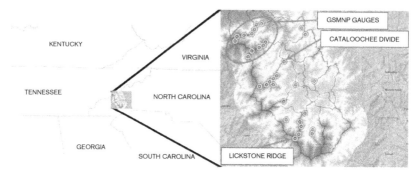

FIGURE 8.1. Map of research area.

Mountain Raingers, worked in partnership with Dr. Ana Barros (Duke University) and her graduate students and postdoctoral researchers.

The day's start time depended on the season and tasks necessary at each gauge site. Most days started early with a long drive into the Great Smoky Mountains National Park (GRSM). The team visited the highest elevation rain gauges first, taking advantage of fresh legs for the most challenging elevation climbs and avoiding pop-up afternoon thunderstorms in the warm season, which can bring dangerous lightning to exposed peaks.

Upon reaching the location, the team checked for rattlesnakes and wasp nests underneath the rain gauge base. Then the Mountain Raingers performed a series of steps to document the state and condition of the site, not unlike measures taken at a crime scene. Since Duke-RGN rainfall observations were necessary to validate National Aeronautics and Space Administration (NASA) and National Oceanic and Atmospheric Administration (NOAA) satellite precipitation-estimation algorithms, one of our team's tasks was to *paint a picture* for NASA and NOAA scientists. We captured images in words and photos, documenting the reliability of the collected data.

Site photographs included those taken of the exterior of the gauge and surroundings, the *mouth* of the gauge, overhanging vegetation, and any debris resting inside the gauge (e.g., leaves, pine needles, dead insects, bird poop). Overhanging vegetation is important because it diminishes rainfall amounts measured by the gauge and any debris resting in the gauge mouth delays the time between rain falling into the gauge and its registration with a tip of the gauge bucket.

Once photographs were taken, we removed the gauge cover and measured any water in the untipped bucket. We also checked the gauge level, ideally indicating the gauge was facing straight up to the sky. After that, we connected the laptop to the data logger, downloaded the logger observations to the laptop, and if necessary, adjusted the date/time of the logger. At the same time, other Mountain Raingers cleaned the gauge cover funnel, mesh, and siphon, allowing the unobstructed flow from raindrop capture to collection in the gauge-tipping bucket. They also cleared a five-foot radius of vegetation and, if needed, trimmed tree limbs. After packing up and one final scan, we headed out. The total visit time at a single gauge location, assuming no glitches, was about forty minutes. On a good day, four rain gauges were visited on each day trek in the GRSM. Over the years, there have been numerous unforeseen challenges adding significant time and decreasing the number of sites visited on a given trek. In every case, the Mountain Raingers got an education beyond the basics of hydrological field research methods.

WHO ARE THE MOUNTAIN RAINGERS?

Undergraduate field research has been recognized as a vehicle to provide deeper, hands-on learning for foundational concepts in the atmospheric sciences (Horel, Ziegenfuss, & Perry, 2013; Quardokus, Lasher-Trapp, & Riggs, 2012). Research opportunities provided for undergraduate students are often controlled to keep project breadth and depth manageable. As such, most research opportunities take place in laboratories on university campuses. For outdoor and off-campus learning experiences, research is still often highly controlled due to logistical constraints. Each research opportunity is valuable for applying concepts and providing a taste of doing, and considering the benefits of a research-oriented career.

On relatively rare occasions, when there is research funding, undergraduates are able to participate in a field project. Most atmospheric sciences field projects are high-intensity, short-duration enterprises, usually through funding from a national agency such as NASA, NOAA, or related agencies. A benefit of undergraduate student participation is gaining an appreciation of logistical challenges related to observing rapidly evolving phenomena, such as thunderstorms, and using optimal positioning of personnel and instrumentation to address project hypotheses.

Occasionally, a field project has long-term funding and offers a remarkable number of learning benefits beyond those already listed. Prolonged projects allow for a longer observation period and extended data collection period that record large-scale and long-term constraints (climatology). The purpose of this chapter is to describe learning benefits for undergraduate students in the atmospheric sciences provided by a specific prolonged field project. This chapter describes the establishment of the high-elevation rain gauge network in partnership with colleagues at the GRSM and Duke University. Additionally, it summarizes key insights from a full-length book in preparation, *Mountain Raingers: An Education Beyond Basic Field Research in the Pigeon River Basin of the Southern Appalachians.*

WE CAN'T PREDICT WHAT WE CAN'T OBSERVE

By September 8, 2004, the remnants of Hurricane Frances had already seriously impacted western North Carolina (Franklin et al., 2006). The village just outside the Biltmore Mansion grounds experienced severe flooding, and several vehicles, homes, and businesses were destroyed along the Swannanoa River, just south of downtown Asheville (National Climatic Data Center, 2004). Numerous landslides were reported in western North Carolina and it was estimated that more than ten inches of rain fell in the mountains during a seventy-two-hour period. Asheville's water supply was compromised when pipes transporting water to the city collapsed after earth under the pipes was scoured away by rapidly flowing floodwaters (North Carolina Department of Environmental Quality, 2017).

The first raindrops of the remnants of Hurricane Ivan began to fall over western North Carolina within nine days of the departure of Frances. The oversaturated ground had little time for recovery and was unable to absorb additional water. Rainfall from Ivan produced instantaneous flooding and additional landslides. The strong sustained and gusty winds downed numerous trees as the soupy soil could offer little resistance to the uprooting trees. Although the total rainfall amounts were less than from Frances, the flooding during Ivan was severe; the outer eastbound lane of I-40 eroded away and collapsed into the raging Pigeon River near the Tennessee and North Carolina border (Schlosser, 2004).

Examination of rain gauge observations made during Frances and Ivan indicated that the heaviest recorded rainfall occurred along the Gulf Coast, Florida, and in the western North Carolina mountains (Barros, 2006). Heavy rainfall over the two coastal locations had been expected, as the storms still packed hurricane-force winds and intensity just before making landfall. However, western North Carolina observations indicated that the southern Appalachian Mountains experienced a nearly 300 percent increase in precipitation for the two tropical storms.

The severity and widespread nature of flooding, tree uprooting, and landslides over western North Carolina was not anticipated. These hazards were driven by precipitation and exacerbated by the mountains. Thus, estimating precipitation over mountainous terrain is an inherently challenging task.

MAKING OBSERVATIONS TO MAKE BETTER PREDICTIONS

Rain gauges are the most accurate method of estimating rainfall. Their primary disadvantage is they can provide observations at only a single location. It is impractical to place rain gauges on every mountaintop of western North Carolina; thus, weather forecasters rely on satellite and radar estimates of precipitation. The challenge of using satellite- and radar-based observations to estimate rainfall is that these instruments do not directly measure precipitation and their observations can be contaminated by substantial ground clutter in the mountains.

Improved engineering and technology has allowed recent satellite and radar instruments to scan thinner layers of the atmosphere to provide, hopefully, more reliable, uncontaminated precipitation data. In anticipation of launching a next-generation satellite instrument in February 2014, NASA funded the Duke University (Dr. Barros) research proposal in 2007 to install and maintain a rain gauge network at high elevations in the western North Carolina mountains. The goal of the project was to provide rainfall observations in the remote GRSM and surrounding region and advance understanding of precipitation processes. Specifically, the study was designed to (a) provide ground validation of precipitation retrieval algorithms, (b) determine measurement uncertainty of satellite-based precipitation estimates, (c) improve understanding of precipitation processes, and (d) assist in the development of next-generation precipitation retrieval algorithms.

The original study, which was funded for a three-year period, was extended by NASA several times and concluded with the Integrated Precipitation and Hydrology Experiment (IPHEx), a field project that took place over western North Carolina (Barros et al., 2014). A follow-on study using the gauge network continued the work started in 2007 and represents a partnership of funding provided by NOAA-National Environmental Satellite, Data, and Information Service (NESDIS), the Center for Western Weather and Water Extremes of the Scripps Institution of Oceanography, and UNCA.

Building the Rain Gauge Network

Locating gauges within the Pigeon River Basin at elevations of at least three thousand feet above sea level was a primary requirement of the Duke-RGN project. Generally, wind speeds are stronger at higher elevations; therefore, rain gauge sites were located just below the primary ridgeline to provide some shielding. Most clouds and rain produced during the warm season in the southern Appalachians blow in from the south. Therefore, a majority of rain gauges were placed on southeast-facing mountainsides. One of the greatest installation challenges was finding sites that had a sky clearing large enough to accommodate a rain gauge in the continuous forest canopy of the Pisgah National Forest and the GRSM.

As approvals for the rain gauge locations were acquired, plans were formulated to haul large and heavy aluminum rain gauges, drainage rocks, fifty-pound bags of cement, and metal posts from vehicle-accessible roads to individual remote locations. The most significant logistical challenge started at the Cosby, Tennessee, campground parking lot in the GRSM, sitting at 2,200 feet. The approved rain gauge site on Mount Guyot was at 6,570 feet. This elevation gain of 4,370 feet took place over a 7.5-mile hike along the Snake Den Ridge Trail and Appalachian Trail (AT). The final 0.3 miles required bushwhacking through thick pine forest to reach the summit.

Fortunately, all trails located near approved gauge locations were accessible by horse. Members of the Great Smoky Mountain Chapter of the Backcountry Horsemen of North Carolina hauled most of the gear and supplies from a road to a shelter or drop-off point along the Mount Sterling Ridge Trail or the AT. From mid-May through mid-June 2009, Mountain Raingers hauled individual loads of drainage rocks, cement powder, and a metal post or a rain gauge from the Cosby Knob backcountry shelter or Mount Sterling drop-off point to individual gauge sites.

On the first day at each location, Mountain Raingers cleared a five-foot radius of vegetation, dug a hole two feet deep, and then placed rocks, cement, and a gauge post in the hole. On the second day, a Mountain Rainger carried a rain gauge from its drop-off point, mounted it to the gauge post, and activated the data logger, making it ready to observe rainfall. Installation of the final gauge of the Duke RGN occurred at Mount Guyot on June 12, 2009.

Travel-Related Logistical Challenges

Although the Duke-RGN was "next door" to UNCA, several factors could derail the best-laid plans for rain gauge visits. Assuming optimal conditions, these all-day visits required driving forty-five to seventy-five minutes to reach the vehicle access point. The remote GRSM gauge locations required a substantial hike; thus a visit was typically a ten- to fourteen-hour investment. Infrastructure support provided by GRSM colleagues was crucial to completing day treks. Interruptions to road and facility access in the GRSM forced by a federal government seques-tration (October 2013), federal position hiring freeze (January–March 2017), and stimulus-related road construction (spring 2010) led to sig-nificant delays in the completion of seasonal gauge visit campaigns.

In the spring, summer, and autumn, the Mountain Raingers made visits to each rain gauge of the Duke-RGN for maintenance and data downloads. Winter conditions proved too challenging to make treks worthwhile. Ten to fourteen hours of daylight were easy to secure dur-ing the summer season, but proved challenging toward the beginning of the spring and toward the end of the autumn.

When arriving for our trek, there was a moment, usually after we shut the vehicle door, when we knew that we were surrendering control of the day. Backpack supplies reflected conditions we expected to face, but uncertainties and unknowns were common. The greatest unknowns, particularly during summer, were the changing weather conditions that we encountered miles and hours away from our vehicle at significantly higher elevations. Balmy, clear, and calm morning conditions could yield to thunderstorms, downpours, and gusty winds by midafternoon. The possibility of hypothermia loomed, even during the hot summer months, above five thousand feet for hikers unprepared with adequate gear. The threat of lightning impacted planning and preparation and the execution of tasks during numerous gauge visit campaigns. Students new to the Mountain Raingers team were required to acclimate them-selves by conquering a series of treks, with increasing intensity, ranging from eight to twelve miles along some of the most scenic trails in the Great Smoky Mountains.

Nature-Related Challenges

Western North Carolina is home to a large population of black bears. An estimated fifteen hundred black bears reside in the GRSM (NPS, 2017a). One of our first surprises in the early days of the Duke-RGN occurred on installation day on Mount Sterling Ridge in May 2009. Sup-plies left by the Backcountry Horsemen, discovered in a state of terrible disarray, alerted us to the first RGN disruption caused by a black bear.

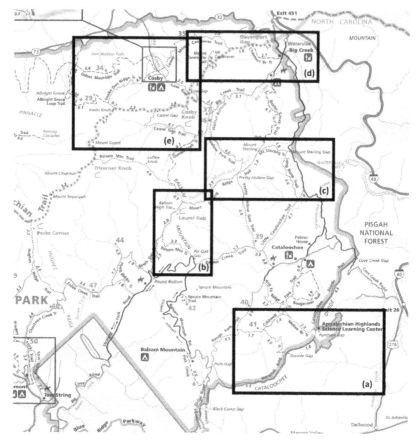

FIGURE 8.2. Map of all the research sites in Great Smoky Mountains National Park. Courtesy of the National Park Service.

A fifty-pound bag of cement had been carried down the mountainside. The initial weight of the bag and its traveled distance indicated the likelihood of our cement thief being a 250-pound adult male bear. Other evidence included a five-gallon plastic water jug with multiple bear claw and teeth punctures. A park ranger later informed us that black bears like plastic chew toys, and coupled with the curious "swishing" sound of the water inside, made our water jug an irresistible challenge.

Another backcountry GRSM challenge are mice and chipmunks, which have learned that humans (and their backpacks) are food sources. It isn't unusual to be awakened in the middle of the night by a rodent scampering across your face or sleeping bag. For this reason, snakes are a welcome inhabitant of the GRSM, as they keep the rodent population in check.

Rain gauges of the Duke-RGN are placed in localized clearings of a generally continuous forest canopy, making them ideal spots for reptile warming. Upon arrival at a rain gauge location, conducting a visual scan of the area around the gauge, looking for sunning reptiles near the base of the gauge post before approaching it, became common practice. The reason for this was due to the *Big Creek Dance* of May 2010. This final gauge visit, on May 26, 2010, at our AT East trek, was located near Big Creek, the lowest elevation gauge of the Duke-RGN (3,398 feet). The late afternoon sun provided direct light on the gauge and surrounding surface. A Mountain Rainger stepped ahead of me, eyes fixed on the gauge. In an instant I noticed movement near the base of the gauge post. I put my hands on the student's shoulders and sauntered, almost dancing, past the gauge to a downslope position. At the same time, a mid-sized timber rattlesnake moved upslope, within a foot of our hiking boots, and settled in a patch of undergrowth five feet from the gauge post. We breathed a sigh of relief as we attended to the gauge, one eye on the instrumentation and another on the snake. Fortunately, timber rattlesnake encounters have been few and uneventful due to the nonaggressive nature of GRSM rattlesnakes (NPS, 2017b).

LESSONS LEARNED: POTENTIAL IMPACTS ON
THE MOUNTAIN RAINGERS

As of July 1, 2017, thirty-nine UNCA undergraduate students have been members of the Mountain Raingers team since the installation of the rain gauge network. A survey of professional and personal benefits of involvement with field research was collected via social media and email, and yielded a 37 percent response rate. The following section summarizes the students' learning related to their involvement in field research of the rain gauge network.

Although all atmospheric sciences students at UNCA are required to complete a Meteorological Instrumentation course before graduating, many do not make a meaningful connection to why field scientists would want to make accurate *in situ* rainfall observations. By providing a link between observing rainfall and validating NASA and NOAA-NESDIS satellite rainfall-estimating algorithms, the Duke-RGN project exemplifies a real-world application for students, deepening their understanding of the importance of measurement accuracy. Additionally, this experience helps them understand the necessity of applying quality control and error estimates to each rainfall observation. They learn that making accurate weather observations is complicated by numerous factors, some of which are entirely out of human control. Furthermore, they learn how to take field measures and minimize errors that *are* in human control (e.g., time drift of a data logger). Finally, they learn the advantages and limitations of *in situ* data in the mountains.

Beyond these somewhat complex ideas are the practical data collection skills that participants learned. These include (a) how to manipulate and display data (observations) through computer procedures and programming; (b) planning a work schedule; (c) organizing data, tools, and personnel; (d) properly siting and installing a rain gauge; (e) physical interaction with a rain gauge to ensure proper functioning; (f) troubleshooting equipment problems in the field; and (g) adjusting plans to maintain safety. In this way, students learned how to plan, organize, and improvise directly through actual experiences. For example, students may have learned that a day's work might be cut short because the laptop battery was not fully charged or work at a gauge was delayed because the hex wrench wasn't in its designated pocket in the tool bag.

Students found that they gained an appreciation for making observations in the "not-so-friendly" confines of the complex and steep terrain of the southern Appalachian Mountains. The rapid weather fluctuations forced them to make wise real-time decisions about adjusting the plan-of-the-day. Students also learned how to deal with exhaustion, when after hours of hiking and climbing, nothing seemed to be going according to the original plan. Patience, a sense of humor, and teamwork are critical in these situations; thus students learned to appreciate the camaraderie of a shared challenge.

However, difficulties and challenges weren't what kept most of the students coming back; a sense of satisfaction and altruism was important to the Mountain Raingers. Several Raingers explained that "knowing at the end . . . that the instrumentation was in better condition than when we'd found it" and "knowing rainfall observations were more valuable (and reliable) due to our contribution to the gauge's upkeep and calibration" made them feel like they were contributing. They felt they were "doing an important service to the scientific research community" and "supporting valuable data collection for a precipitation-wise tomorrow." In a few cases, students decided that field work was not consistent with their interests and respectfully declined continued participation in the project upon completion of their first seasonal visit.

Mountain Raingers expressed, with near unanimity, that participation in the Duke-RGN project opened their eyes to opportunities not previously considered as a career in meteorology and climatology. This included work as a research scientist, specifically working with instruments and observations in the field. Three project graduates found jobs that include field work and several others are in graduate school, preparing for similar types of jobs. Although a rigorous assessment methodology was not applied, we hypothesized that student involvement in the Duke-RGN-oriented field research made it more likely that Mountain Raingers would pursue postbaccalaureate education by attending graduate school ($13/31 \times 100\% = 41.9\%$) compared to the general atmospheric sciences undergraduate student population ($45/164 \times 100\% = 27.4\%$) at UNCA. Our collaboration with Duke helped to expand our undergraduate students' horizons by allowing them to witness firsthand research conducted at a research-intensive university, using our rainfall observations to aid in investigations of their hydrological hypotheses. One Mountain Rainger moved directly into a PhD program at Duke after finishing work on the project as a UNCA undergraduate. Another outcome of note was some students gained the confidence to explore less familiar disciplines within meteorology, including scientific communication.

CONCLUSION

Most Mountain Rainger graduates are no longer involved with field research, but noted the important personal revelations that impact who they are today. This was a result of having learned how to push themselves mentally and physically to overcome preconceived limits and challenges, reach lofty goals and adventures, and pace themselves. Finally, non-career-related personal benefits of participation included deeper connections with classmates, professors, nature, and the outdoors. They discovered increased confidence to explore the mountains (sometimes alone) during leisure time, the knowledge that it is possible to combine the things you love (backpacking and weather), and unforgettable views. A student summed it up, saying this project "kept me connected to the reason I started to become interested in the science in the first place."

Acknowledgments

The author acknowledges the NPS, UNCA, and Duke University students who have assisted with this project since 2007. Financial support was provided by NASA grants NNX07AK40G, NNX10AH66G, and NNX13AH39G. The author also thanks Dr. Ana Barros and her research group at Duke University through funding provided by the Pratt School of Engineering at Duke University. Funding for maintenance support of the Duke-RGN was provided by F.M. Ralph at Scripps Institution of Oceanography, NOAA-NESDIS NC-CICS grant 2014–2918–08, and an undergraduate research grant from UNCA.

References

Barros, A. (2006). *Measuring, modeling, and understanding orographic precipitation regimes and hydrology in mid-latitude mountain regions.* A proposal submitted to and funded by NASA. NASA project grant NNX07AK40G.

Barros, A.P., Petersen, W., Schwaller, M., Cifelli, R., Mahoney, K., Peters-Liddard, C., . . . Zipser, E. (2014). *NASA GPM-ground validation: Integrated precipitation and hydrology experiment 2014 science plan.* Retrieved from http://dx.doi.org/10.7924/G8CC0XMR

Franklin, J.L., Pasch, R.J., Avila, L.A., Beven, J.L., Lawrence, M.B., Stewart, S.R., & Blake, E.S. (2006). Atlantic hurricane season of 2004. *Monthly Weather Review, 134,* 981–1025.

Horel, J.D., Ziegenfuss, D., & Perry, K.D. (2013). Transforming an atmospheric science curriculum to meet students' needs. *Bulletin of the American Meteorological Society, 94,* 475–484.

National Park Service. (2017a). Black bears: Great Smoky Mountains National Park. Retrieved from https://www.nps.gov/grsm/learn/nature/black-bears.htm
———. (2017b). Reptiles: Great Smoky Mountains National Park. Retrieved from https://www.nps.gov/grsm/learn/nature/reptiles.htm

National Climatic Data Center. (2004). *Storm data* [Data file]. Retrieved from https://www1.ncdc.noaa.gov/pub/orders/IPS/IPS-50F0D405-ACA4-4899-9C06-81B0F895A233.pdf

North Carolina Department of Environmental Quality. (2017). Recognizing landslides. Retrieved from https://deq.nc.gov/about/divisions/energy-mineral-land-resources/north-carolina-geological-survey/geologic-hazards/landslide-types-causes

Quardokus, K., Lasher-Trapp, S., & Riggs, E. (2012). A successful introduction of authentic research early in an undergraduate atmospheric science program. *Bulletin of the American Meteorological Society, 93,* 1641–1649.

Schlosser, J. (2004, September 16). Portions of I-40 closed indefinitely, DOT says. *NC News & Record (Greensboro, NC).* Retrieved from http://www.greensboro.com/news/general_assignment/portions-of-i—closed-indefinitely-dot-says/article_7bf9a713-b48f-545c-a27f-0736b88b806a.html

The back of people paddling a canoe at Grand Teton National Park. Courtesy of the National Park Service.

Health and Self

Empowering Learning in Parks

This section is about the power of parks to facilitate learning about ourselves. Some of the most important lessons we face as a society are rooted in our ability to understand our own mental and physical states. Whether for emotional, environmental, or physical learning, or just to see ourselves as learners, parks provide empowering spaces. This collection of chapters illustrates how we "find ourselves" in place and how our own diversity connects and empowers us to discover and learn more about ourselves and our country.

From where does our love of nature come? McMillan poses this question in her chapter "Learning Environmental Psychology in the National Parks." She explains how psychological experiences of the natural world shape environmental attitudes and values. Environmental psychology provides tools to facilitate connections with the natural world, experience its benefits, and ponder the value of these experiences. This chapter explains, explores, and reflects upon a college-level course taught at Rocky Mountain National Park. The course demonstrates key tenets and benefits of environmental psychology—both in practice and as an area of study.

Increasingly, research shows that national parks and natural areas provide visitors with ideal locations for engaging in healthy physical activities. Focusing on human health and well-being and as part of the National Park Services' (NPS's) Health Parks Healthy People program, this team of researchers set out to see if educational messaging could make a difference in promoting healthy physical behaviors among visitors. In "Can Signage Influence Healthy Behavior? The Case of Catoctin Mountain National Park," Bose, Nagle, Benfield, Costigan, Wimpey, and Taff investigate this question about signs at this Maryland national park. The authors detail the results of their exploratory experiment and wrestle with whether signage, which is a feasible and easily employed strategy, actually makes a difference.

Sometimes learning about ourselves means understanding our heritage. View and Guiden invite us to see what happens when an NPS ranger invites sixteen racially and ethnically diverse middle school students from a public school in Washington, DC, to go on a tour of twelve Civil War and civil rights sites. In the chapter "Learning Historic Places with Diverse Populations: An Exploratory Study of Student Perceptions," we learn how personal encounters with historical sites coupled with interpretation can impact student retention of historical content.

Across the country and across disciplines, the field trip, as we now know it, is an endangered or at least a "threatened" species. In geoscience education, especially at the college level, it is simply not feasible to take students to many, if not most, of our national parks. In the chapter "'I Felt Like a Scientist!' Accessing America's National Parks on Every Campus," Bursztyn, Goode, and McDonough introduce virtual and augmented reality field trips as a potential solution. The Grand Canyon Expedition app was designed to facilitate virtual field trips, but the authors found that it also facilitates student empowerment. The authors adopted a pedagogical perspective on mobile learning from Kearney et al. (2012) and used it as a framework to understand three key aspects of mobile learning in the Grand Canyon Expedition app: personalization, collaboration, and authenticity. Tools like this app help learners see themselves as scientists, while creating opportunities for enhanced adult learning, conservation, and citizenship more broadly.

Standing on frozen Emerald Lake at Rocky Mountain National Park. Courtesy of Kent Deardorff.

9

Learning Environmental Psychology in the National Parks

DONNA K. MCMILLAN

INTRODUCTION

Throughout our evolutionary history, humanity has been embedded in the natural world. The biophilia hypothesis proposes that, because of this history, we come equipped with an innate tendency to be responsive to nature and life processes (Wilson, 1984). However, many who love the national parks also cite impactful experiences from childhood or as adults, even if at the time they may not have been fully aware of the effects of the experience. As an eight-year-old from West Virginia, when my family took a two-week vacation that included visits to several western national parks, did I know the seeds that were getting planted? Inspection of my "travel journal" from that time is not particularly revealing, although clearly I was interested in the wildlife: "Today my mommy saw a moose." But even now memories linger: the awe-inspiring Front Range of the Colorado Rockies rising before us as we approached from the east, the strong sulfur stench of Yellowstone's geysers, the fantastical red hoodoos of Bryce Canyon, the excitement of riding horses near the Grand Canyon. It turns out that these landscapes spoke deeply to me, although I neither realized nor could have articulated it. I was not consciously aware of a longing to return, although each year for over twenty years, my husband and I have headed west to these iconic national parks—to camp, hike, and be among their wonders. Our most frequent and best-known destination is Rocky Mountain National Park.

So then, from where does love of nature come? This question, about our psychological experience of the natural world, is a focus of research in environmental psychology, an area that can be taught in the national parks. Like most psychological phenomena, connection to nature is likely influenced by both genetic inheritance and personal experience. In this chapter, I explain, explore, and reflect upon a college-level course I teach that demonstrates the key tenets and benefits of environmental psychology—both in practice and as an academic discipline.

WHAT CAN WE LEARN IN THE NATIONAL PARKS?

The idea of exploring the *psychological* significance of the natural world may seem surprising. More typically, people recognize our national parks as premier venues for experiential learning about stories of American peoples, places, and natural resources. Typical programming in the parks, from hour-long ranger-led evening programs to day-long classes, often focuses on topics like cultural history, ecology, wildlife biology, natural history, and geology, as well as artistic activities such as photography, painting, and nature writing. However, the national parks also have rich potential for visitors to explore their personal *experience* of the natural world and to think about and reevaluate the role and meaning of nature in their lives.

Indeed, this is the focus of an off-campus undergraduate college course, Environmental Psychology at Rocky Mountain National Park (RMNP), which I have taught six times over the past fourteen years (McMillan, 2012). A regular, full-credit college course, it explores the psychological significance of our national parks and investigates ways in which the natural world affects people psychologically. Although our college is located in Minnesota, we are able to offer the course in Colorado because our academic calendar includes a January term during which students take one intensive course, which may be off campus. Thus, for the Environmental Psychology course, in January, a group of twenty to twenty-two students and I live at the YMCA of the Rockies campus adjacent to RMNP.

The course is intentionally academic and experiential, and students are encouraged to explore and engage with the national park—on hikes during the day or in the full moonlight, while snowshoeing, stargazing, sitting and observing, and watching the abundant wildlife. We also meet for two- to three-hour class sessions each weekday to read and discuss relevant literature, theory, and research. Our experience is enhanced dramatically by interactions with national park rangers, when they meet with us for class discussions or lead us in an outing, such as a winter ecology snowshoe hike. The integration of academic theory and research with students' immediate experience in the national park spans three disciplines: psychology, environmental studies, and American studies.

THE ENVIRONMENTAL PSYCHOLOGY AT RMNP COURSE

Although the subfield of environmental psychology generally addresses interactions of humans with many types of environments, including social and built environments, the Environmental Psychology at RMNP course focuses specifically on our relationship to the natural environment. This is an area within environmental psychology sometimes called conservation psychology (Clayton & Myers, 2015) or ecopsychology (Kahn & Hasbach, 2012). In recent years, the psychological effects of nature on humans has been getting more attention in the scholarly literature (e.g., Clayton & Myers, 2015) and in popular writing (e.g., Louv, 2008; Williams, 2017).

Environmental Psychology at RMNP focuses particularly on four areas: (a) subjective experience in nature, (b) empirical research about people's relationships with the natural environment, (c) cultural factors, and (d) nature-related systems and institutions—particularly the national park system. To address these four core areas, students engage in a variety of types of learning, as outlined in table 9.1.

Not only is the *what* (i.e., the content) of the course important, but also the *how*. While the course includes some brief lectures, it is primarily a discussion-based seminar, encouraging students to actively think about and share their thoughts about course material. To ensure that the experiential aspects of the course do not come to feel like mandatory daily assignments, students are given much autonomy to encounter and engage with nature. This practice maximizes students' intrinsic motivation and allows for the powerful effects that unstructured, unscripted interaction with nature can have (see Louv, 2008). Students value the sense of exploration and discovery and overwhelmingly praise this aspect of the course. "The ability to choose which activity we did many afternoons was a highlight for me," wrote one student. Another explained in detail that

TABLE 9.1   TYPES OF LEARNING AT ROCKY MOUNTAIN NATIONAL PARK

Core areas	Types of learning	Examples
(a) **Subjective experience**	Explore the national park. Spend time every day intentionally developing a more intimate relationship with the natural world.	In "learning expeditions" students investigate and write about self-chosen aspects of the natural environment and share findings with the group.
(b) **Empirical research**	Read and evaluate empirical research to learn about ways in which nature is psychologically significant.	Topics include the following: • Stress reduction theory (Bratman, Daily, Levy, & Gross, 2015), which explores physiological indicators of stress in natural environments • Cognitive, interpersonal, emotional, and psychological benefits of time in nature, such as attention restoration and enhanced creativity (Kaplan, 1995), increased prosociality (Zhang, Piff, Iyer, Koleva, & Keltner, 2014), enhanced positive affect, increased vitality (Ryan et al., 2010), and contemplative, self-reflective, and spiritual experiences
(c) **Cultural factors**	Explore nature's role in one's sense of self and identity. Investigate human connections to other animals, wild or domesticated. Learn a bit about how nature can be a helpful part of psychotherapy. Examine humans' effects on the environment, with a focus on attitudes and behaviors related to sustainability (e.g., Harré, 2012).	We spend a good deal of time discussing how the natural world fits into modern American lives, given factors such as our ever-increasing urbanization, busyness, and use of technologies. We ask questions like "To what extent do people today have interaction with and knowledge of the natural world? And what are the implications?"
(d) **Systems & institutions**	Learn about the National Park Service. Explore why the national parks are managed as they are.	Students explore different facets of the National Park Service. The Organic Act of August 25, 1916, states that "the Service thus established shall . . . conserve the scenery and the natural and historic objects and the wildlife therein and provide for the enjoyment of the same in such manner and by such means as will leave them unimpaired for the enjoyment of future generations." A more affective facet offers the perspective that a park such as Rocky Mountain National Park is "essentially a symbol of nature and its pace and power," where one can find that "nature, taken on its own terms, has something to say that you will be glad to hear" (Sax, 1980, pp. 12, 15).

the ability to go on excursions by ourselves, and the amount of autonomy we were given allowed us to reflect on what we were learning in very personal ways. . . . For example, we learned about what makes a place wild—and I remember reading the point that said "opportunities for solitude" and immediately feeling a connection between the environment in the Rockies and that statement. We were given so many opportunities to seek out solitude and silence in nature, and it reinforced what we were studying, the benefits of spending time in the wild.

Anonymous student feedback has been very positive and consistently reveals that students find the course valuable and love experiencing the national park in a sustained, intentional manner: "One strength [of the course] definitely is being able to get out of the classroom and live what we are learning about." Some students reflect explicitly on the importance of the course being held at RMNP, and they describe bidirectional effects. In other words, they claim that the course helped them to experience the park in more profound ways, and the park helped them learn more in the course:

> Studying at the national park definitely enhanced my learning experience in the environmental psychology class. The experience allowed me . . . to experience the things that we were learning about at the same time that we were learning them. Learning about the effects that nature experiences could have on us gave more meaning to being in the national park and changed the way I will think when in any other national park.

Another student wrote, "I loved being in the national park, and I learned so much about the National Park System, how it works, how it is affected by political decisions, and what park rangers do. It made all the difference in the world to study here." A third student reported,

> I do not think that I would have had the same mindset of experiencing nature as *intentionally* as I did had we not been in a setting like RMNP. The opportunity to study in a place so widely recognized for its natural beauty and peacefulness made me want to experience it more, and made it easier to be aware of nature's ability to calm and satisfy psychological needs—i.e., topics we discussed in class!

A fourth student emphasized the broad-ranging impacts: "I loved being able to experience and reflect on the course material in such a powerful environment. This course touched all aspects of my life."

The last quote touches on the contemplative and self-reflective experiences that many students reported. It is clear that the course and the setting together prompt significant exploration of both outer and inner terrains. One student explained, "I think it was an exploratory month which provided spaces for students to open up to the world around them and get to know themselves more." Another student reported, "Our readings came alive by being in the place where they were set. Plus, I found the month to be an interesting case study into my own psychological well-being. Taking the time to reflect on my ups and downs, my highs and lows was very valuable." Furthermore, gaining insight into their own relationship with nature can be a highlight for students:

> Before this class I thought that I had a pretty good grasp on why the environment is important to our well-being, but boy, I really had no idea! From every book to discussion to assignment to adventure in nature I learned something new every day and had my own previous conceptions challenged. I think my favorite aspect of the class was learning about something and then experiencing it in nature or thinking back to different experiences I had already previously had. I understand my own love for the natural world so much better now.

REFLECTIONS AND LESSONS LEARNED

The success of this course may be due in part to the extent that the course experiences—particularly the time spent in nature—support psychological well-being. A leading area of research on psychological well-being, *self-determination theory* (SDT; see Ryan & Deci, 2017), focuses on the value of intrinsic motivation (where individuals are motivated by their own interests or values rather than by external rewards or pressures) and the central importance of three psychological needs: autonomy, relatedness, and competence. A large body of international research suggests that when these psychological needs are met, individuals are more likely to thrive and to experience intrinsic motivation and psychological well-being, including better mood, vitality, self-esteem, and physical health (Ryan & Deci, 2000). Environmental Psychology at RMNP, by design, nourishes these three psychological needs. The autonomy built into the course allows students' nature engagement to remain intrinsically motivated, and thus they get more out of the experience than if these were mandated, scripted outings.

Regarding the psychological need of relatedness, when I first started teaching the course, I put much thought into how to establish good group dynamics and foster closeness among group members. While I still attend to this issue, I have found that the nature setting itself facilitates relationships. This is reinforced in what a student wrote: "Never have I connected to people so quickly or intimately or on so many different levels as I did in Colorado." Furthermore, while self-determination theory conceptualizes relatedness as close, personal relationships with other people, relatedness can also happen with regard to the natural world. Research suggests that a sense of connectedness to nature can have similarly powerful effects (Mayer, Frantz, Bruehlman-Senecal, & Dolliver, 2009).

The third psychological need in the SDT model is competence, and indeed students seem to gain feelings of competence throughout the course. These energetic, college-aged individuals are eager to get out and test themselves. Many of them are clearly motivated by a desire for mastery, wanting to challenge themselves in various ways and to succeed. Many express pride in completing the various hikes, from easier nearby terrain, to scaling pathless Eagle Cliff, to a hardy few who choose to climb 12,324-foot Flattop Mountain. They write of challenges faced, obstacles overcome, and psychological growth experienced.

While the students' nature encounters are sometimes motivated by mastery, other motivations can be present as well. Kellert (2012), in his exploration of the biophilia hypothesis, proposed a variety of ways that people relate to nature, including through (a) attraction/aesthetic pleasure, (b) curiosity and reason, (c) exploration, (d) aversion and fear, (e) practical material uses, (f) affection and emotional connection, (g) mastery and dominion, (h) spirituality, and (i) symbolism. Individuals differ regarding which of these motivations are most salient. In students' choices of how to interact with nature, they often are following the types of connection that are most appealing or familiar to them. It is important to allow for flexible expression of these individual variations. It can also be helpful, however, to point out these other ways of relating to nature, potentially opening students up to new types of experiences.

THE NATIONAL PARKS AND MODERN LIFE

Psychologically, national parks are perhaps more important now than ever, providing a counterbalance to a variety of tendencies of modern life. While an anthropocentric view of nature (i.e., valuing nature because of its usefulness to humans) runs the risk of blinding us to the inherent value of nature (a perspective often described as ecocentric), it is clear that natural connection can provide many benefits. The sublime nature of a place like RMNP can move us out of our usual patterns, helping us to consider how we live our lives. In such a place, ideally we encounter the wild rather than the urban; we experience a slower pace of life influenced by nature's rhythms; we find some freedom from technology, multitasking, consumerism, and other human artifacts; and our senses are awakened as we remove our earbuds and listen to the environment. We become more physically active than sedentary; our involuntary attention is activated while our effortful, directed attention rests; we experience quiet rather than noise; we encounter flora and fauna; and we meet up with forces bigger than ourselves. It can be a relief to realize that not everything is about us. Years ago, trailhead signs at RMNP included the statement "The mountains don't care," a straightforward reminder that helps to counter some anthropocentric tendencies of modern life.

As explored in depth in Environmental Psychology at RMNP, research shows that engagement with nature can have a range of positive psychological effects. Even virtual nature, in the form of photographs or videos, has been shown to have such effects. However, effects generally are stronger in real nature settings, and some nature settings are more powerful than others.

For most Americans, nearby nature is not an iconic national park with sublime landscapes. As Williams (2017) suggests, we can think of encounters with nature as varying by dose, from daily encounters with smaller, nearby environments, to weekly outings in a larger natural area, to grand time spent in the national parks. Our national parks could be the largest, most powerful dose of nature available, and the effects of experiences there could continue to inspire and motivate long afterward. Perhaps experiences in such areas can even sensitize us to noticing, appreciating, and seeking out more common, less spectacular nature. For some students in Environmental Psychology at RMNP, it also sparks a desire to return to RMNP by working in a summer job nearby, by just visiting, or even by becoming an interpretive ranger.

CONCLUSION

We can all benefit from becoming more educated about and consciously aware of the psychological benefits of nature experience. While most will not get the opportunity to have a month-long course focused on such benefits at a national park, we can still benefit from messages that encourage us to be intentional and mindful in our experience of nature.

Early experiences outdoors can plant the seeds of later devotion to the natural world. But perhaps it is never too late. Recently, a small group of us were casually talking around a campfire in Chaco Culture National Historical Park when a park ranger remarked on the value of dark skies. Recalling the first time that he saw the Milky Way, he described the experience as "life changing." When did this happen for him? Surprisingly, it was not a childhood experience, but occurred after he had become a ranger at Chaco Canyon. There, in the darkness of an International Dark Sky Park, he experienced the wonder of observing the night sky, and was transformed.

For me, connection to nature was nourished by my childhood family vacation to the western national parks. My professional work in this area started with my conscious recognition of the psychological importance of nature in my own life and the enormous benefit of time spent in our natural parks. Whether watching elk and moose among the Colorado River willows of RMNP's Kawuneeche Valley or having the rare experience of seeing a mountain lion as it crossed the road one January near Hollowell Park in RMNP, my life, too, is transformed by experiences in the national parks. Not so long ago, humanity dwelled more intimately in the natural world; today, while we enjoy the many benefits of modern life, we sometimes need reminders of the value and necessity of nature. The study of environmental psychology provides tools to facilitate intentional connections with the natural world, experience its benefits, and ponder the value of these experiences in our lives.

References

Bratman, G.N., Daily, G.C., Levy, B.J., & Gross, J.J. (2015). The benefits of nature experience: Improved affect and cognition. *Landscape and Urban Planning, 138*, 41–50.

Clayton, S., & Myers, G. (2015). *Conservation psychology: Understanding and promoting human care for nature* (2nd ed.). Hoboken, NJ: Wiley-Blackwell.

Harré, N. (2012). *Psychology for a better world: Strategies to inspire sustainability*. Auckland, New Zealand: University of Auckland.

Kahn, P.H., & Hasbach, P.H. (2012). *Ecopsychology: Science, totems, and the technological species*. Cambridge, MA: MIT Press.

Kaplan, S. (1995). The restorative benefits of nature: Toward an integrative framework. *Journal of Environmental Psychology, 15*, 169–182.

Kellert, S.R. (2012). *Birthright: People and nature in the modern world*. New Haven, CT: Yale University Press.

Louv, R. (2008). *Last child in the woods: Saving our children from nature-deficit disorder*. Chapel Hill, NC: Algonquin Books.

Mayer, F.S., Frantz, C.M., Bruehlman-Senecal, E., & Dolliver, K. (2009). Why is nature beneficial? The role of connectedness to nature. *Environment and Behavior, 41*, 607–643.

McMillan, D.K. (2012). Environmental psychology at Rocky Mountain National Park: An undergraduate academic and experiential course. *Ecopsychology, 4*, 102–109.

Ryan, R. M., & Deci, E. L. (2000). Self-determination theory and the facilitation of intrinsic motivation, social development, and well-being. *American Psychologist, 55,* 68–78.

———. (2017). *Self-determination theory: Basic psychological needs in motivation, development, and wellness.* New York, NY: Guilford Press.

Ryan, R. M., Weinstein, N., Bernstein, J., Brown, K. W., Mistretta, L., & Gagné, M. (2010). Vitalizing effects of being outdoors and in nature. *Journal of Environmental Psychology, 30,* 159–168.

Sax, J. (1980). *Mountains without handrails: Reflections on the national parks.* Ann Arbor: University of Michigan Press.

Williams, F. (2017). *The nature fix: Why nature makes us happier, healthier, and more creative.* New York, NY: W. W. Norton.

Wilson, E. O. (1984). *Biophilia.* Cambridge, MA: Harvard University Press.

Zhang, J. W., Piff, P. K., Iyer, R., Koleva, S., & Keltner, D. (2014). An occasion for unselfing: Beautiful nature leads to prosociality. *Journal of Environmental Psychology, 37,* 61–72.

Walker looking at sign about healthy park visit at CATO. Courtesy of Heather Costigan.

10

Can Signage Influence Healthy Behavior?

The Case of Catoctin Mountain National Park

MALLIKA BOSE, LARA NAGLE, JACOB BENFIELD,
HEATHER COSTIGAN, JEREMY WIMPEY,
AND B. DERRICK TAFF

INTRODUCTION

With more than eighty-five million acres across 419 individual units, US National Park Service (NPS) units provide many opportunities to improve human health and well-being through the natural, cultural, and educational resources provided in these unique and rare protected areas. With upwards of 331 million yearly visitors from all over the world, these areas have the potential to provide myriad health benefits for visitors. Such benefits can be optimized through targeted educational and communicative strategies (NPS, 2003; Roggenbuck, Loomis, & Dagostino, 1990). Furthermore, increasing human perceived value of the environment promotes support for the conservation and rehabilitation of natural areas that in turn increase health resiliency (Aronson, Blatt, & Aronson, 2016).

In the past several decades, sedentary behavior has increased, resulting in lower rates of physical activity that are in part to blame for the rapidly rising levels of chronic disease in the United States (e.g., obesity, cardiovascular disease, type 2 diabetes; Bai, Stanis, Kaczynski, & Besenyi, 2013). In America, the obesity rate is over 22 percent in every state (State of Obesity, 2017). Natural protected areas provide visitors with an ideal location for engaging in healthy physical activities, which is a common motivation for visitation (Godbey, Mowen, & Ashburn, 2010). Physical activity is generally associated with positive health outcomes that can aid in reducing chronic health symptoms (e.g., lower blood pressure, lower risk of diabetes, lower risk of osteoporosis, lower risk of certain cancers; Ben-Shlomo & Kuh, 2002).

Programs using nature for therapeutic, stress reducing, and/or physical therapy benefits have shown great promise in mitigating symptoms related to medical afflictions (Annerstedt & Währborg, 2011). Daily contact with nature or visiting a park on a daily basis has been shown to improve health and well-being in both adults and children (Maller, Townsend, Pryor, Brown, & St. Leger, 2006). In addition, natural protected areas such as national parks offer free or low-cost opportunities for health benefits, making the benefits more accessible to a broader audience (Godbey et al., 2010).

NATIONAL PARK SERVICE HEALTHY PARKS HEALTHY PEOPLE RESEARCH STUDY

Initiatives such as the NPS Healthy Parks Healthy People Program (HPHP), launched in 2011, formally recognize the countless health benefits provided by natural areas and parks (NPS, 2013). As part of HPHP, research partners study NPS units and other types of protected areas to assess the impact of specific park resources on the health and well-being of visitors (Maller et al., 2009; Thomsen, Powell, & Allen, 2013). Prior to this research, no studies had focused on increasing physical activity in park units using cost-effective educational strategies such as signage. Educational messaging through signage may be a feasible and easily employed strategy for influencing visitors to increase physical activity during their park experience.

Research indicates that educational messaging is most effective when it is clear and succinct, and occurs early in the visitor's planning process (Cole, Hammond, & McCool, 1997; Ham, 2016). Educational strategies are also more effective when they encourage personal responsibility and significance (Knapp & Forist, 2014), are contextually precise (Vagias & Powell, 2010), and are from a trustworthy and dependable source (Manfredo & Bright, 1991). Additionally, placement of messages— often in the form of signage in or adjacent to an area where a given behavior is preferred—can be effective in influencing visitor behaviors (Davis & Thompson, 2011; Hall, Ham, & Lackey, 2010; Hockett & Hall, 2007).

In this study, researchers considered the influence of educational signage on hiking behavior at Catoctin Mountain National Park (CATO).

In addition, the study examined how many visitors indicated that they noticed the treatment educational signage, and of those who noticed the educational signage, what their self-reported behavior was as a result. In terms of programming, researchers assessed visitors' awareness of the HPHP initiative, how important visitors think it is for the NPS to promote parks as health resources, and how best to do so.

METHODS

Site Description and Sampling

CATO is set in a heavily wooded area, with mountainous, rocky outcroppings and picturesque streams. The park is well known as a hiking destination, within close proximity to the greater Washington, DC, area. The park is also the home of Camp David, the presidential retreat established in the 1940s under the Franklin D. Roosevelt administration. Although the retreat is not accessible to the general public, a nearby trail system is open for public use. There are five trails that offer hiking lengths of approximately 1.6 kilometers, four trails offering lengths ranging from 4 to 8 kilometers, and one 12.9-kilometer perimeter loop trail. Additionally, the park hosts camping, historic cabins, and picnic areas. Together, these amenities, set within a scenic location, draw over 220,000 visitors annually (NPS, n.d.).

Sampling for this study took place at the CATO Visitor Center during August 2016 (one of the busiest months for visitation). The Visitor Center is located on the main thoroughfare to and from the park at a junction offering access to the interior sections of the park, and hosts full-time NPS interpretive employees, park displays, maps, and general information about the area.

Treatment and Control

An HPHP educational message was crafted with the assistance of CATO staff to ensure mission relevancy and to be as concise as possible, while triggering personal significance through contextually specific language. A green sign with white font suggesting that visitors "Experience Catoctin Mountain's beautiful streams, hardwood forests, and vistas while improving your health; consider hiking further this visit . . ." was used as the treatment. The treatment messages were centrally placed in and outside the CATO Visitor Center (see chapter-opening photo).

Data Collection

The signage was deployed for eight days during the treatment sampling period. Control days, with no HPHP-installed signs, occurred over a separate eight-day sampling period. Visitors were approached to participate in an anonymous and voluntary prehike survey. Participants were then given a Garmin eTrex 10 GPS unit to record their hiking activity along the route. Upon their return, visitors turned in the GPS unit and completed the posthike survey. For the analysis, surveys were coded and linked to participant GPS data. A total of 121 park visitors received GPS units, with a participation acceptance rate of 83 percent; however, data cleaning resulted in 111 usable tracks—53 for control days and 58 for treatment days.

FINDINGS

Demographic and Self-Reported Health Characteristics

Similar to the average demographic profile of NPS visitors (Manning, Anderson, & Pettengill, 2017), the study participants were mostly white, educated, and middle-aged. The control and treatment groups were similar to one another in terms of race, gender, age, and education level. Participants in the GPS study also provided self-reported assessments of personal health through the survey. On both treatment and control days, more than 90 percent of all participants ranked their quality of life and physical, mental, and social health as good, very good, or excellent.

Effects of Signage

To examine the influence of the treatment signage as a form of education to increase health-related behaviors, the researchers compared five quantitative variables representing hiking activity for control and treatment groups: (a) total time spent hiking, (b) total distance hiked, (c) cumulative ascent, (d) cumulative descent, and (e) average speed. A t-test comparing the control and treatment groups revealed no significant difference in any of these five variables.

To understand the possible causes for this lack of difference in hiking behavior between control and treatment groups, the researchers examined whether the study participants in the treatment group (i.e., those exposed to signage encouraging healthy behavior) remembered seeing the signage. Responses to the postsurvey question *Did you notice any park signage or information encouraging you to be more physically active and healthy during your visit?* were evaluated by treatment and control groups. There was a slight increase in the number of survey respondents who said *yes* to this question during a treatment day. However, the overall majority of respondents on both control and treatment days (88.6% and 77.8%, respectively) did not report seeing the HPHP signage.

In other words, the presence of signage did not ensure that the sign and its message registered with viewers or those passing by. Interestingly, 11 percent of visitors in the control group mentioned that they had seen park signage that encouraged them to be more physically active. It is likely that these visitors were referring to permanent CATO signage and/or informational material available at the Visitor Center.

Open-ended survey responses indicated that for a small sample of respondents ($n = 15$), the treatment signs and educational messaging were effective in influencing healthy physical behaviors. For example, one survey respondent reported that signs encouraged him or her to "walk uphill." Another participant reported that the messaging encouraged him or her to "hike farther," and an additional visitor reported that signage encouraged him or her to "walk longer by choosing another trail." For another respondent, seeing an educational message made them "more excited . . . so [he or she] moved faster," while another indicated that the signage generally "inspire[d] [them] to be more healthy." Thus educational signage prompted healthy behavior in a subset of park visitors.

Awareness and Promotion of HPHP and Parks as Health Resources

Subsequently, the researchers explored visitor perceptions of parks as health resources through the promotion of the HPHP initiative. Most participants, across the control and treatment groups ($n = 89$), indicated that they supported the parks as health resources (< 93%) despite being unfamiliar with the HPHP initiative (< 90%).

In terms of how to promote the HPHP initiative, open-ended survey responses ($n = 39$) indicated that educational information about HPHP should be shared primarily online and via social media. Visitor centers ranked second on the list of places where HPHP health resources and programming should be promoted, followed by TV and radio, schools, and urban areas, respectively.

CONCLUSIONS

This study suggests that HPHP education has an opportunity to improve upon and expand beyond signage in parks to realize the program goals. The study found that the educational message used for the treatment did not have a significant influence on visitors' hiking activity. Only about 22 percent of the participants in the study noticed the signage during the treatment condition. However, qualitative assessments from participants who noticed signage indicated that educational messaging did encourage them to hike farther or more quickly. More research is needed to better understand the placement, wording, and design of signage that optimizes healthy physical behavior, for example, novel and motivating signage located at challenging spots along the hiking route.

Researchers may consider treatments designed with consideration to "stealth health" approaches (Evans & Hastings, 2008; McCuaig & Quennerstedt, 2018). These ultimately encourage healthy behavior while not explicitly stating that the action is healthy. Future research should also attempt to collect a larger sample over a much longer period of time to help determine if there is a ceiling effect on visitor behavior in CATO. For example, encouraging visitors to hike farther in this setting may be irrelevant, perhaps due to a limited number of hiking trails or because a hiking saturation point is already being reached. Visitors may also predetermine their physical activity goals prior to arriving at the park.

Moreover, if parks are to be recognized and used as health resources, they must be accessible to groups with the highest incidence of obesity and chronic diseases. In the United States, health risks are disproportionately borne by minority and low socioeconomic status groups (National Academies of Sciences, Engineering, and Medicine, 2017; State of Obesity, 2017). Communication about the physical health benefits of parks should be shared directly with these audiences. Study participants suggested primarily using communication outlets such as online and social media sites to promote HPHP programming, followed by visitor centers, TV/radio, schools, and venues in urban areas. It should be noted that social media and online platforms, which have the advantage of wide and diverse viewership, could be leveraged to encourage park visitation in addition to healthy behaviors while visiting parks.

In addition, emerging technologies now allow for in-situ educational deliveries, such as mobile apps (Mihanyar, Rahman, & Aminudin, 2016). Recent research approaches that have employed real-time ambulatory strategies—delivering educational messaging and collecting data through devices such as smart phones, tablets, and watches—may be effective in promoting and examining health-related behaviors (Heron & Smyth, 2010).

Acknowledgments

The authors would like to thank CATO and other NPS staff, including Diana Allen and Vincent Santucci, for their assistance. This research could not have taken place without the collective input from the Penn State University HPHP Working Group and the help of Sohrab Rahimi. We also thank the Penn State Institutes for Energy and the Environment for funding.

References

Annerstedt, M., & Währborg, P. (2011). Nature-assisted therapy: Systematic review of controlled and observational studies. *Scandinavian Journal of Public Health, 39*(4), 371–388.

Aronson, J., Blatt, C., & Aronson, T. (2016). Restoring ecosystem health to improve human health and well-being: Physicians and restoration ecologists unite in a common cause. *Ecology and Society, 21*(4), 39.

Bai, H., Stanis, S. A. W., Kaczynski, A. T., & Besenyi, G. M. (2013). Perceptions of neighborhood park quality: Associations with physical activity and body mass index. *Annals of Behavioral Medicine, 45*(1), 39–48.

Ben-Shlomo, Y., & Kuh, D. (2002). A life course approach to chronic disease epidemiology: Conceptual models, empirical challenges and interdisciplinary perspectives. *International Journal of Epidemiology, 31*, 285–293.

Cole, D. N., Hammond, T., & McCool, S. (1997). Information quality and communication effectiveness: Low-impact messages on wilderness trailside bulletin boards. *Leisure Sciences, 19*(1), 59–72.

Davis, S. K., & Thompson, J. L. (2011). Investigating the impact of interpretive signs at neighborhood natural areas. *Journal of Interpretation Research, 16*(2), 55–68.

Evans, W. D., & Hastings, G. (2008). Public health branding: Recognition, promise, and delivery of healthy lifestyles. In W. D. Evans & G. Hastings (Eds.), *Public health branding: Applying marketing for social change* (pp. 3–24). Oxford, UK: Oxford University Press.

Godbey, G., Mowen, A., & Ashburn, V. A. (2010). *The benefits of physical activity provided by park and recreation services: The scientific evidence.* Ashburn, VA: National Recreation and Park Association.

Hall, T. E., Ham, S. H., & Lackey, B. K. (2010). Comparative evaluation of the attention capture and holding power of novel signs aimed at park visitors. *Journal of Interpretation Research, 15*(1), 15.

Ham, S. (2016). *Interpretation: Making a difference on purpose.* Golden, CO: Fulcrum.

Heron, K. E., & Smyth, J. M. (2010). Ecological momentary interventions: Incorporating mobile technology into psychosocial and health behaviour treatments. *British Journal of Health Psychology, 15*(1), 1–39.

Hockett, K. S., & Hall, T. E. (2007). The effect of moral and fear appeals on park visitors' beliefs about feeding wildlife. *Journal of Interpretation Research, 12*(1), 4–26.

Knapp, D., & Forist, B. (2014). A new interpretive pedagogy. *Journal of Interpretation Research, 19*(1), 33–38.

Maller, C., Townsend, M., Pryor, A., Brown, P., & St. Leger, L. (2006). Healthy nature healthy people: "Contact with nature" as an upstream health promotion intervention for populations. *Health Promotion International, 21*(1), 45–54.

Maller, C., Townsend, M., St. Leger, L., Henderson-Wilson, C., Pryor, A., Prosser, L., & Moore, M. (2009). Healthy Parks Healthy People: The health benefits of contact with nature in a park context. *The George Wright Forum, 26*(2), 51–83.

Manfredo, M. J., & Bright, A. D. (1991). A model for assessing the effects of communication on recreationists. *Journal of Leisure Research, 23*(1), 1–20.

Manning, R. E., Anderson, L. E., & Pettengill, P. (2017). *Managing outdoor recreation: Case studies in the national parks* (2nd ed.). Wallingford, UK: CAB International.

McCuaig, L., & Quennerstedt, M. (2018). Health by stealth: Exploring the socio-cultural dimensions of salutogenesis for sport, health and physical education research. *Sport, Education and Society, 23*(2), 111–122. doi:10.1080/13573322.2016.1151779

Mihanyar, P., Rahman, S. A., & Aminudin, N. (2016). The effect of national park mobile apps on national park behavioral intention: Taman Negara National Park. *Procedia Economics and Finance, 37*, 324–330. doi:10.1016/S2212-5671(16)30132-0

National Academies of Sciences, Engineering, and Medicine. (2017). *Communities in action: Pathways to health equity.* Washington, DC: National Academies Press. doi:10.17226/24624

National Park Service. (2003). Reviewing our education mission: Report to the National Leadership Council. Retrieved from https://www.nps.gov/archeology/aiassess/2003Educ_Mission.pdf

———. (2013). The national parks and public health: A NPS Healthy Parks, Healthy People science plan. Retrieved from https://www.nps.gov/subjects/healthandsafety/upload/HPHP_Science-Plan_2013_final.pdf

———. (n.d.). National Park Service visitor use statistics. Retrieved from https://irma.nps.gov/stats/

Roggenbuck, J. W., Loomis, R. J., & Dagostino, J. (1990). The learning benefits of leisure. *Journal of Leisure Research, 22*(2), 112–124.

State of Obesity. (2017). Obesity rates and trends. Retrieved from http://stateofobesity.org/rates/

Thomsen, J. M., Powell, R. B., & Allen, D. (2013). Park health resources: Benefits, values, and implications. *Park Science, 30*(2), 30–36.

Vagias, W. M., & Powell, R. B. (2010). Backcountry visitors' leave no trace attitudes. *International Journal of Wilderness, 16*(3), 21–27.

Dexter Avenue Baptist Church. Courtesy of George Mason University.

11

Learning Historic Places with Diverse Populations

An Exploratory Study of Student Perceptions

JENICE L. VIEW AND ANDREA GUIDEN

INTRODUCTION

What happens when a National Park Service (NPS) ranger invites sixteen racially and ethnically diverse middle school students from a public school in the Washington, DC, area to go on a tour of twelve "Civil War to Civil Rights" sites? How do personal encounters with historical sites and professional interpreters impact student retention of historical content beyond the typical measures of standardized tests (View & Azevedo, 2015)? We set out to understand the impact of the Freedom Rangers adventure on a diverse student population, and we used the frameworks of place-based education and interpretation to guide our inquiry and analysis.

Funded by a grant from the NPS Centennial fund, an NPS ranger, a classroom teacher, and an embedded researcher created a yearlong after-school "Junior Rangers" program. It was designed to engage students with historical content through visits to NPS sites within the region and reflections beyond. Joined by a parent volunteer (and twenty students and chaperones from another nearby eighth-grade group), the Freedom Rangers set off, in May 2016, on a two-thousand-mile, ten-day bus journey to visit a dozen sites across the Southeast, as listed in table 11.1.

Site number	NPS site name & location	NPS content & activities
1	Tredegar Iron Works Richmond, VA	Civil War churches, cemeteries, hospitals, prisons Civil War battlefields Workers (variety) at the Iron Works City of Richmond
2	Appomattox Court House Appomattox, VA	The Civil War surrender site, terms of the surrender Generals Grant & Lee The McLeans home (where surrender was signed) African American contributions
3	Cowpens National Battlefield Chesnee, SC	Turning point in Revolutionary War Women and African American contributions Details of the Battle of Cowpens
4	Dr. Martin Luther King, Jr. National Historic Site Atlanta, GA	Sweet Auburn neighborhood Ebenezer Baptist Church The Dr. Martin Luther King Center Dr. Martin Luther King birth home Timeline of Dr. Martin Luther King's life
5	Tuskegee Institute National Historic Site Tuskegee, AL	Tuskegee Institute & buildings The Oaks (home of Booker T. Washington) Milbank Agricultural Building & George Washington Carver research site
6	Tuskegee Airmen National Historic Site Tuskegee, AL	Tuskegee Airmen Moten Field (Hangars I & II)
7	Selma to Montgomery Historic Trail Selma, AL	54-mile trail with 11 specific landmarks 10 sidebar cards re specific people, events, & organizations
8	Little Rock Central High School Little Rock, AR	The US Constitution, segregation/resistance Arkansas resistance US Supreme Court cases Central High School building Bullying of the Little Rock Nine
9	National Civil Rights Museum Memphis, TN	32 exhibits, including slavery, Jim Crow, civil rights battles/resistance in the South, Dr. Martin Luther King, Black Power, Dr. Martin Luther King assassination, etc.

10	Stones River National Battlefield Rutherford County, TN	Battle of Stones River Union victory Emancipation Proclamation Cemetery
11	Andrew Johnson National Historic Site Greeneville, TN	Early life/home Presidency Impeachment Cemetery
12	Booker T. Washington National Historic Site Hardy, VA	Washington's birthplace *Up from Slavery* Education Hampton University Tuskegee Institute

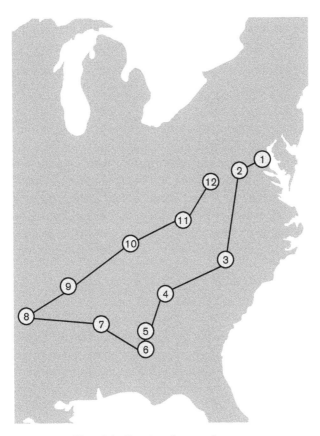

FIGURE 11.1. Map of the Freedom Ranger sites.

PLACE-BASED EDUCATION

The interpretation of historic sites focuses on bringing the historical significance of a place to life. To bring sites to life for students in the Freedom Rangers program, we applied many of the principles of place-based education to these culturally specific, human-created sites. There are five distinct types of place-based educational programs found in North American schools: (a) cultural studies programs (e.g., oral histories, journalism, dramatic plays), (b) local nature studies, (c) real-world problem solving, (d) internships and work experiences, and (e) the community process of local decision-making (Wattchow & Brown, 2011). In an extensive study of ten programs representing more than one hundred schools in twelve states, the Place-Based Education Evaluation Collaborative (2010) found that place-based education fosters students' connection to place and creates vibrant partnerships between schools and communities. It boosts student achievement and improves environmental, social, and economic vitality. Advocates for place-based education argue that

> educational biodiversity falls prey to the bulldozers of standardization. . . . More and more, we drive a wedge between our children and the tangible beauty of the real world. (Sobel, 2004, p. 4)

Furthermore, Furman and Gruenewald (2004) assert that through critical, place-based pedagogies, democratic action research begins when children and youth start investigating their own familiar places, identifying issues, analyzing them, and then planning and implementing some sort of action. As award-winning high school history teacher James Percoco (see chapter 3, "Invoking the Spirit of History") reminds us, "Every community has a history. Every community has a story and those stories are invariably connected to that national narrative of American history" (p. 33).

INTERPRETATION

Freeman Tilden articulated six principles of interpreting national parks and historic sites. For our purpose, we were interested in effective interpretation for children as well as how skilled interpreters can convey to children the "keynote" (Tilden, 2007, p. 93) of an historic place in order to "keep it from seeming to have been frozen at a moment of time when nobody was at home" (p. 102). Yet, interpreters who treat their discipline as something detached from the school-based learners' larger educational journey and remain insulated from broader educational debate and discussions about curricula, pedagogy, and transdisciplinary work do so to the field's detriment (Wattchow & Brown, 2011).

METHODOLOGY

Context

The Freedom Rangers program included three 8th graders, nine 7th graders, and four 6th graders; seven boys and nine girls; three white students, one Latinx, and twelve African American students. One of the white students was a native Portuguese speaker who was attending a US school for the first time.

Students visited the twelve sites, sometimes more than one site per day, over the course of a ten-day field trip. At most sites, students watched an introductory video, visited self-directed exhibits, and then engaged with an interpreter for up to forty-five minutes. The exception was the National Civil Rights Museum, a non-NPS site where the exhibits were entirely self-directed. We joined the students as participant-observers, scribbling field notes and trying to capture the utterances of students, interpreters, and other adults, occasionally posing clarifying questions.

Data Collection

We, as researchers, gathered field notes, student reflections, videos, student interviews, and field conversations among the adult participants to illustrate this place-based learning case study. Our evidence offers a profile of student engagement as a factor in students' attachment to history and retention of historical content. Qualitative analyses of these data revealed four primary themes: (a) student engagement, (b) ranger skill, (c) historical content retention, and (d) prior knowledge.

We collected the written student feedback on the day of each visit. Students reflected on what they (a) learned at the site, (b) still wanted to learn, and (c) wanted to share with others upon returning home; in addition, they outlined (d) how the site related to the modern civil rights movement, and (e) their thoughts about the interpreter/ranger. Some posttrip feedback was videotaped.

Seven months after the trip, during the after-school program, we surveyed the students about their Freedom Rangers experience. The survey posed three questions:

- Of the (listed) sites that we visited, which did you enjoy? (check all that apply)

- Of the sites you checked, select your favorite and write two sentences about why it was your favorite.
- What do you think are the top three characteristics (list of nine adjectives) of a good ranger/interpreter?

These nine adjectives came directly from the responses provided on student feedback forms, which included (a) enthusiastic, (b) loud enough to hear, (c) knowledgeable, (d) asks good questions, (e) welcomes my questions, (f) interactive, (g) passionate, (h) fun, (i) nice, and (j) other (with an explanation).

Ten months after the trip, we conducted interviews with the student participants. Students were shown a list of the sites and were asked to recall everything they remembered, with little prompting or affirmation. They were also asked what they remembered about the interpretation at each site. The interviews were audio recorded for accuracy and transcribed.

Limitations

The scope of the data was limited, thus instead of breadth we offer some in-depth insights. A specific limitation is that we were able to compile longitudinal data for only nine students. The three eighth graders had gone on to high school, one student returned to Brazil, and three completed surveys but were unwilling to participate in the interviews. Despite these gaps, we believe our analysis offers nuance and insight into place-based learning and interpretation with diverse populations of middle school–age students.

For the purpose of analysis, we defined *engagement* through an examination of student questions and comments in the field notes, as well as students' listing or uttering the site as a "favorite."

Ranger skill was formed by adjectives that students used to describe the characteristics of a good ranger. Interestingly, the students generated adjectives that were strikingly similar to Freeman Tilden's framework. Tilden (2007) noted that interpretation should

- be relatable,
- inspire revelation,
- demonstrate artistry,
- offer provocation,

- share the whole rather than the parts, and
- be tailored to the audience. In other words, that children's interpretation is different, although not a dilution of what is offered to adults.

Historical content included the information contained in NPS museum brochures, NPS Junior Ranger booklets, and the ranger commentary on-site.

Retention was the sum of utterances or content about each category. If a student remembered more than 50 percent of what the NPS determined was important for Junior Rangers to remember, it was coded as "high retention."

Prior knowledge required a more detailed definition. Sometimes it was direct, explicit information to which the Freedom Rangers had been exposed. Generally, prior knowledge was indicated by student expression on-site or during an interview. It is possible that students had prior knowledge of some sites (such as Little Rock Central High School or the National Civil Rights Museum) and did not express it in the on-site feedback or interview. However, the only situation where some prior knowledge was assumed for every student was for Dr. Martin Luther King, Jr., as many of the students had visited the national memorial in Washington, DC.

FINDINGS

Data from five of the sites—Tuskegee Airmen, Selma to Montgomery, the Dr. King site, Little Rock Central High School, and the National Civil Rights Museum—indicated high retention of historical content. Of these, Tuskegee Airmen, Selma to Montgomery, the Dr. King site, and Little Rock were identified as favorite sites seven months later.

Little Rock Central High School stood out as the site with the highest combined rating seven months later. Five of the nine students remembered it favorably, three indicated that it was their favorite site, and one student indicated that the ranger was their favorite. At the ten-month mark, two students rated the Little Rock ranger a favorite, the highest for any. Yet, the retention of historical content for Little Rock (five students) was not as high as for Selma (eight students), Tuskegee Airmen (eight students), or the Dr. King site (six students).

The Freedom Rangers' retention of historical content seemed to be influenced by prior knowledge and the ranger, or engagement with the site. Within the top five sites, the students had prior knowledge of each, except for the Tuskegee Airmen site. These same five, minus the Selma site, generated mentions of the ranger. For seven of the twelve sites, many students expressed no memory of the historical content they were exposed to ten months earlier, and none had prior knowledge, even though for three sites, students remembered liking the ranger.

We were also curious about the sites where there was evidence of indifference or dislike of a ranger, such as at the Dr. King site and the Tuskegee Airmen site. The site for which the students had the greatest amount of prior knowledge—the Dr. King site—was not a favorite among the students, and they found the ranger to be lacking in the characteristics they valued. On the flip side, one male student gave insight to why the Tuskegee Airmen site ranked so high despite no prior knowledge and indifference to the ranger: "[It] had all the activities that we could interact with; it is important to me to be able to touch things; interactive things help me to figure things out more, helps me to understand more about them." Regarding the Selma site, where no one remembered the ranger, one of the students expressed a deep emotional connection with the marchers during Bloody Sunday, saying "I actually got to walk in their footsteps [over the Edmund Pettus Bridge], where they walked, where they got hurt, where . . . some people died; I got to walk on it; the [local] cars did not slow down to take in the moment, I guess it's cuz they live there; since I didn't live there I would have slowed down to take in the moment of crossing the bridge."

We also looked into student reactions at the National Civil Rights Museum, the one site that lacked an interpreter. At this site, students went on a self-guided exploration, and while three female students expressed a preference for this way of learning, one noted that "if there was a ranger, I could've asked questions that the writings didn't answer." We can only speculate about the impact of the ranger at Little Rock in terms of historical content retention and the highest student engagement (three favorites). Perhaps it is related to the ranger having been the overall favorite at the time of the trip.

Based on our analyses, we propose four key insights about student learning at historic sites. Table 11.2 expands upon these insights.

Insight 1: **Prior knowledge** yields higher levels of engagement.

Insight 2: Guided interpretation is useful for helping **students engage** with the site.

Theme	Description	Examples of evidence (from field notes, on-site feedback sheets, surveys, and/or interviews)
Prior knowledge	Any expressions that the historical content was familiar prior to the trip	W girl: The scene on the bridge in the movie *Selma* was shot in Selma. W girl: [It was my favorite because] I did my project on that and had to present [on it]. (Little Rock) AA girl: Civil Rights Museum in Memphis was my favorite because I like how it was formatted almost like a maze. I also liked that I finally got to see the famous hotel.
Student engagement	Utterances or written statements about "favorite" site	Latina: My favorite was going to Selma to Montgomery National Historic Trail because I got to walk where many black people lost and fought for their life. I loved the movie *Selma*, so I was happy to visit the site that it all happened in. . . . [It was] a big experience for me, I was crying a little because it felt like happy that I got to walk across that bridge. I was so happy. W girl: [Civil Rights Museum was my favorite] because a lot more things including a recreation of the bus that Rosa Parks took, and of the bus that the Freedom Riders took, it might have been the actual bus and it included a lot of different things we did not know about, like the Black Panthers. AA boy: My favorite site was the Little Rock High School. That was my favorite because we had a deep conversation of what went on during this time.
Ranger skill	Enthusiastic Loud enough to hear Knowledgeable Asks good questions Welcomes my questions Interactive Passionate Fun Nice	African immigrant girl: (Cowpens) They both were really active, they knew how to make us laugh. . . . They were nice and they knew how to handle children. There are people who will treat you like you're five, but you might be thirteen or treat you like an adult. . . . You have limitations. They understand, just by looking at us they know how we are. AA girl: The ranger got distracted kind of easily. I don't know if was because of the noise or what, but I really believe that she got distracted easily. Also there were times when I couldn't hear her. AA boy: [The ranger] was all about audience participation.

TABLE 11.2 *(Continued)*

Theme	Description	Examples of evidence (from field notes, on-site feedback sheets, surveys, and/or interviews)
Historical content retention	Determined by National Park Service rangers and materials	Latina: The Tredegar Museum showed different places where the Civil War took place, and the Civil War was all about helping slaves. Today we still have civil rights going on, and the Civil War helped to build into, for example, "Black Lives Matter." W boy: MLK [sic], his father, and his grandfather were all pastors at the Ebenezer Baptist Church. This is important because this means that MLK's family was totally church-born and followed the Lord with his heart. AA girl: [Booker T. Washington was] born in slavery, but did more in his life than most people. He had a school on the road type thing. He wanted to educate himself and others.

NOTE: W = white; AA = African American.

Insight 3: **Ranger skill** impacts the level of student engagement and enjoyment.

Insight 4: Greater engagement leads to **students' retention** of information.

DISCUSSION

This qualitative case study draws from a body of data that included more detail on each student (disaggregated by race, language, gender, and grade level) and further analysis of each site's interpretation and the longitudinal impact on each student. In some cases, the adults on the trip had a different perspective of the site than the students and engaged in discussions on the bus or at meal time to learn more about student perspectives. In several cases while traveling, the multiracial group of students debated vigorously among themselves about nonviolence as a tactic of social change versus armed resistance (particularly at the Dr. King site and the National Civil Rights Museum), the pace of social change over the centuries (multiple sites), the illogic of racial segregation (especially at Stones River Cemetery), and the balance of power between the federal and state governments (at Little Rock Central High School). This engagement was notable.

However, the study intended to gauge the longer term retention of historical content and, to a lesser extent, enjoyable memories of visiting NPS sites. No student said that the trip was a negative experience or a waste of time. In some cases, students had good memories of the site or rangers, but no memory of the historical importance of the site or the content. For example, many students remembered that there was a scavenger hunt at Tredegar Iron Works and that there was a battle reenactment at Cowpens Battlefield without remembering why. Similarly, many students confused the Tuskegee Institute and the Tuskegee Airmen field site during their interviews. Finally, there were cases where the student seemed to be having an "off" day at the time of the interview or would have remembered more specific historical content with visual prompting, but could only conjure weather-related or behavioral memories ("it was very hot" or "we went to the gift shop" or "we sang 'we shall overcome'"). In short, the evidence offered only suggestions of impact.

CONCLUSIONS

Most middle school social studies classrooms, textbooks, and standardized tests, and therefore many teachers, promote rather flat stories of US history—ones that lack nuance, controversy, or multiple perspectives (Swalwell, Pellegrino, & View, 2015). Classroom conversations about African Americans and social justice tend to be limited to very narrow opportunities, such as Black History Month and the Dr. Martin Luther King holiday. The ten-day Freedom Rangers trip, on the other hand, provoked multiple and wide-ranging conversations about US racial justice struggles.

This was an unusual trip due to many factors—including the funding source, the design (as part of a yearlong after-school program), the multi-age grouping, and the in-depth experience of the chaperones. Additionally, our team was not typical of field trip chaperones. Nevertheless, the evidence seems to point to a positive relationship between student engagement and an historic site. There appeared to be value in having the site interpreted by a professional. Student retention of historical content ten months after a site visit was evidence of the depth of this engagement. There seem to be several critical elements for helping a diverse group of students find personal connections with national park historic sites. These include building intentional relationships between teachers and rangers, paying more attention to the development of student prior knowledge by classroom teachers and rangers, and engaging in age-appropriate interpretation by rangers on-site. We would do well to integrate these elements into future programs.

References

Furman, G. C., & Gruenewald, D. A. (2004). Expanding the landscape of social justice: A critical ecological analysis. *Educational Administration Quarterly,* 40(1), 49–78.

Place-Based Education Evaluation Collaborative. (2010). *The benefits of place-based education: A report from the Place-based Education Evaluation Collaborative* (2nd ed.). Retrieved from http://tinyurl.com/PEECBrochure

Sobel, D. (2004). *Place-based education: Connecting classroom and community.* Great Barrington, MA: Orion Society.

Swalwell, K., Pellegrino, A., & View, J. L. (2015). Teachers' curricular choices when teaching histories of oppressed people: Capturing the U.S. Civil Rights Movement. *Journal of Social Studies Research,* 39(2), 79–94.

Tilden, F. (2007). *Interpreting our heritage.* Chapel Hill: University of North Carolina Press.

View, J. L., & Azevedo, P. C. (2015, November/December). Attempting to teach and interpret beyond standardized tests: Learning historic places with diverse populations. *Legacy Magazine: National Association of Interpretation,* 26(6), 33–35.

Wattchow, B., & Brown, M. (2011). *A pedagogy of place: Outdoor education for a changing world.* Clayton, Victoria, Australia: Monash University.

Students on the campus of California State University, Fullerton explore the Grand Canyon with an interactive learning app. Courtesy of Natalie Bursztyn.

12

"I Felt Like a Scientist!"

Accessing America's National Parks on Every Campus

NATALIE BURSZTYN, RICHARD GOODE, AND COLLEEN MCDONOUGH

INTRODUCTION

Imagine . . .

It's one-quarter of the way through the semester and the students still have their game faces on, but the nonmajors are clearly just trying to get through their science requirement by memorizing key terms.

The first lecture on geologic time was on Monday. Now it's Wednesday, and ninety-five students are lined up outside the classroom door with their smartphones, Snapchatting, texting, and Facebooking their friends about how boring yet another lecture is going to be, especially on such a perfect, warm fall day.

Today, however, will be different.

After the instructor opens the door, the students shuffle in, head to their usual seats, and make no move toward opening their notebooks or putting their phones away. The instructor interrupts their reverie to inquire how many students have smartphones or tablets. More than three-quarters of the class does.

"Okay," she says, "get with a partner so there's at least one smartphone or tablet between the two of you." The students shuffle, rustle, and murmur amongst themselves, eventually forming partnerships. She says, "While you were choosing partners, I emailed the class an app. Please go ahead and install it now, then we'll head outside." The students fiddle with their phones, suddenly intrigued.

Soon, there are ninety-five college students roaming around the campus quad. They animatedly discuss which direction is east and what "superposition" means again, so they can beat their classmates to the next location on their virtual field trip to the Grand Canyon.

The instructor overhears her students asking each other questions: "What was the difference between 'dis-' and 'non-' conformity again?" "Where did you find the youngest unit at that Hance Rapid stop?" "What was your score at the end?" "Did you get a helicopter ride?"

Now, instead of trying to memorize a dozen new vocabulary terms, the students are reviewing the unit on stratigraphic principles in a physical and applied practical manner. The students are informally competing against each other to earn the highest score so they can get a virtual helicopter ride out of Grand Canyon and win the game.

This case study describes the learning facilitated by the Grand Canyon Expedition App (GCX), an augmented reality field trip tool for college-level geology courses.

NATIONAL PARKS, FIELD TRIPS, AND GEOLOGY

Geomorphology is the study of Earth's landforms and understanding the processes that made them. Long before the first national park was named and the National Park Service Organic Act of 1916 was passed, geoscientists were exploring, sketching, and making scientific hypotheses about the iconic landscapes that would become America's national parks. Probably the most famous of these historic natural philosophers is John Muir. While he was an important figure in the context of the establishment of the National Park Service (NPS), he was also the originator of the glacial hypothesis for Yosemite Valley's origin. Muir proposed that it was glacial processes that worked to carve out Yosemite's U-shaped valleys, waterfalls, and kettle lakes over unimaginable lengths of geologic time, not faults, as had previously been posited. Meanwhile, Clarence Dutton, who was working for the US Geological Survey, interpreted the storybook layers of vast geologic time in Grand Canyon, followed, more famously, by John Wesley Powell, who navigated the Colorado River through the Grand Canyon, inferring the impact of rock types on the width and gradient of those turbulent waters.

The NPS preserves and provides access to these landforms and biomes. For example, the unique biomes within the national parks teach us about endangered and invasive species, coral reefs, frozen tundra, and deserts; the dynamics of predator and prey relationships; and the extreme climates of Death Valley and Denali. Earth and life science educators teach concepts embedded in these unique landscapes. These include those that were carved by the processes of wind, water, and ice; the violent internal process of plate tectonics with active, dormant, and extinct volcanic structures; millions of years of a changing climate; and the recent evidence of anthropogenic climate change.

Today geoscience educators use national park landscapes as examples when teaching new learners of geology the processes that work to shape the Earth's surface. In geoscience education, the field trip, as we now know it, is endangered. Furthermore, due to geographic restrictions, it is simply not feasible to take students to many, if not most, of our national parks.

Field trips have long been a standard in geoscience education. In fact, they drive geoscience culture and are a necessary technique used to augment learning and increase student engagement (e.g., Fuller, 2006; Kastens et al., 2009; Mogk & Goodwin, 2012; Orion & Hofstein, 1994; Tal, 2001). Typically, field trips supplement the lecture and lab components of a geoscience course and provide high-impact experiential learning for the students (e.g., Hovorka & Wolf, 2009; Prince, 2004; Race, 1993). Current trends of increasing class sizes and budget restrictions in higher education are contributing to a decrease in field trip opportunities for introductory-level geoscience students (e.g., Bandiera, Larcinese, & Rasul, 2010; Cook, Phillips, & Holden, 2006; Stainfield, Fisher, Ford, & Solem, 2000; Whitmeyer & Mogk, 2013). In addition, university-sponsored field experiences are often beyond the reach of high-enrollment introductory classes due to liability and time constraints (Friess, Oliver, Quak, & Lau, 2016; McGreen & Sánchez, 2005). The best field sites may be too remote, unrealistic for an online class, or inaccessible due to rough terrain for persons with disabilities and limited mobility. Many introductory geoscience courses are taught using traditional lecture and teacher-centered approaches that may not stimulate interest in geology (Andresen, Boud, & Cohen, 1996; Deslauriers, Schelew, & Wieman, 2011; Mazur, 2009). Thus, there are many opportunities to improve and innovate the introductory geoscience course so that students are inspired to learn more about the physical world.

AUGMENTED REALITY AND VIRTUAL FIELD TRIPS

Everywhere you look, mobile smart devices are permeating our society. Ownership of these devices has grown exponentially among US adults, as have the educational apps to use with them (Anderson, 2015; Dahlstrom & Bichsel, 2014). As a result, student interest and engagement can be positively impacted through the use of portable devices for education (Bursztyn, Shelton, Walker, & Pederson, 2017).

In recent years the development and use of virtual field trips (VFTs) has extended beyond simply showing videos for use in science, technology, engineering, and mathematics (STEM) postsecondary education. Virtual field trips, used mostly in a passive/observational video or computer-based format as alternatives for accessible accommodation or complements to field excursions, have been around for some time and have been increasing in use as technology becomes more accessible and affordable (Beetham & Sharpe, 2013; Friess et al., 2016; Hill, Nelson, France, & Woodland, 2012; Kay, 2012). Studies comparing results from verbal-based and imagery-based tests discovered that students who participated in the imagery-based tests found them more enjoyable and had greater retention of the presented material (Habraken, 1996; Rigney & Lutz, 1976; Tuckey, Selvaratnam, & Bradley, 1991). Physical models are superior to images because students no longer have to try to imagine the object in 3D space. Virtual and augmented reality field trips are the intermediary step between physical models and the real world. In the context of these field trips, virtual reality involves replicating a real environment or creating an imaginary one, and augmented reality involves integrating the real-world environment with supplementary computer-generated elements.

Assessment of the educational impact of VFTs has begun in a number of STEM fields at the secondary level, and VFTs are increasingly being used at the postsecondary level in biology, medicine, engineering, geography, and geology courses (e.g., Friess et al., 2016; Lee, 2012; Liarokapis et al., 2004; Yuen, Yaoyuneyong, & Johnson, 2011). Results from these and other studies indicate that students enjoy the VFTs, and compared with traditional learning, researchers see a gain in interest in the material through the immersive experience (Bursztyn, Shelton, et al., 2017; Friess et al., 2016; Jacobson, Militello, & Baveye, 2009; Pringle, 2013; Spicer & Stratford, 2001; Stumpf, Douglass, & Dorn, 2008).

Recent advances in mobile technology benefit students by enabling them to visually and physically interact with virtual representations of the real world. For example, undergraduate biology students' responses to a VFT on tide pools were overwhelmingly positive (Spicer & Stratford, 2001). The majority of students stated that they had learned a lot, enjoyed using the VFT, and would appreciate seeing more VFTs incorporated into their education. Stumpf et al. (2008) compared the results of students who attended a real-world geomorphology field trip to the results of students who attended a virtual field trip. Those who participated in the virtual field trip performed similarly on concept comprehension to those who participated in the real field trip. Jacobson et al. (2009) constructed a virtual field trip to Mexico's Chinampas Basin and found that students reacted positively to their virtual field trip experience. The students enjoyed the interactivity and became more engaged than they anticipated. Thus, creating more image-heavy, interactive, immersive, geoscience educational tools that students can use with their accessible devices shows enormous promise (Ozdemir, 2010; Pringle, 2014; Spicer & Stratford, 2001).

A Framework for Assessing Mobile Learning

Despite the widespread use of mobile devices, the use of mobile technology as a pedagogical tool remains relatively unexamined from the learner's perspective (Kearney, Schuck, Burden, & Aubusson, 2012). Using the pedagogical perspective on mobile learning (m-learning) from Kearney et al. (2012) as a framework, we are able to identify distinct patterns of positive outcomes from the experiences of students participating in VFTs. The framework emphasizes three significant qualities of m-learning that occur within the time and place of the activity: *personalization, collaboration*, and *authenticity*. This structure helps to explain and situate student responses to experiencing VFTs within current research on learner experiences with m-learning.

Personalization. Learners have some degree of control over the experience and are able to personalize their experience in a number of ways, including (a) using a personal device such as an iPad or a smartphone, (b) controlling the physical space and time in which they complete the activity, and, (c) of particular importance to our study, setting their own goals (Kearney et al., 2012). Combined, these aspects of m-learning lead to an increased sense of ownership in the learning process (Traxler, 2007). Personalized goal setting encourages competition, which can be fostered by learners and their peers. Participants can compete with each other as well as with the educational "game."

Collaboration. Students working together is a vital aspect of the educational experience in which learners make meaning through conversational exchange (Kearney et al., 2012; Sharples, Taylor, & Vavoula, 2007). With m-learning, the use of mobile devices provides a medium where our students can experience collaboration with others as well as with the environment (Kearney et al., 2012).

Authenticity. Students need to see that the experience has real-life context and personal meaning. Kearney et al. (2012) define *authenticity* as the extent to which activities are realistic and offer solutions to problems encountered in the real world. When properly executed, m-learning environments provide accurate, high-level authenticity not found in traditional classroom settings. Students are able to connect with and learn from the activities in realistic virtual spaces that serve as academic "practice fields" (Kearney et al., 2012).

CASE STUDY: THE GRAND CANYON EXPEDITION (GCX) APPS

Three augmented reality (AR) field trips called Grand Canyon Expedition (GCX) were created using smartphone GPS technology. GCX is AR rather than virtual because the user moves around in the real world while virtually navigating Grand Canyon using their smart device. These AR field trips (ARFT) can be used to teach introductory geology concepts in geologic time, geologic structures, and hydrologic processes. In the GCX apps, students complete a series of identification and comprehension questions about geological concepts using examples at different locations within Grand Canyon. Answering questions correctly allows the player to progress down the Colorado River to the next location and exercise. Answering questions incorrectly triggers a pop-up explanation of the answer the student selected and the player is returned to the question and remaining answers. Each time a question is answered incorrectly the number of points possible is reduced, thus reducing the player's total score. This feature provides the opportunity for the player to beat their previous score by replaying the game and answering the questions again (correctly) with fewer attempts. The other type of question in this platform requires the student to examine a photograph and identify a specific geologic feature by tapping or swiping a feature on the touchscreen. This type of question requires observation and critical thinking skills about the content and its application in the real world example of Grand Canyon.

In terms of curriculum, the GCX apps cover introductory units on geologic time, geologic structures, and hydrologic processes as typically

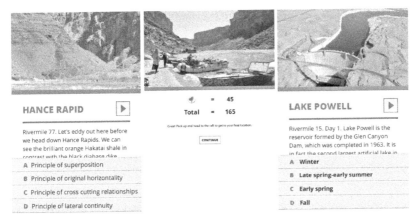

FIGURE 12.1. Screenshots of the Grand Canyon Expedition app.

taught in physical geology, Earth system science, and historical geology courses. Curriculum standards established by the Academic Senate of California Community Colleges, the Academic Senate of California State University, the Academic Senate of the University of California, and the Association of Independent California Colleges and Universities, as mandated by Senate Bill 1440 (Student Transfer Reform Act, 2010), were used as a baseline for student learning objectives. In addition to addressing curriculum at the postsecondary level, the Next Generation Science Standards (NGSS) has addressed this content throughout grades three to twelve.

Do the GCX Apps Increase Student Engagement and Learning?

We initially tested the three GCX apps at five postsecondary institutions with 874 student participants. We found that the apps have a resoundingly positive impact on student *interest* in learning geology (Bursztyn, Shelton, et al., 2017). We also found that the apps had a negligible, but positive, impact on student *learning* of geoscience concepts (Bursztyn, Walker, Shelton, & Pederson, 2017). These results are comparable to numerous other studies involving VFTs, wherein students report having more fun and being more interested, but show no statistical improvement in their content learning.

Results showed that the students who completed at least two of our GCX apps were significantly more interested in learning the geosciences than students who completed only one or none of the AR modules. More comprehensively, results from statistical modeling indicated three strong predictors of student interest in learning the geosciences: (a) initial interest, (b) being a STEM major, and (c) the number of AR field trips completed. Gender and ethnicity had no statistical impact on the results, suggesting the field trip modules have broad reach across student demographics. AR field trips and VFTs for mobile devices therefore have the potential to be an accessible and financially viable means to bring field trips to diverse students who would otherwise experience none.

Applying Kearney's Mobile Learning Framework

We wanted to further assess the GCX AR field trips using key concepts from Kearney's m-learning framework. We used in-depth, qualitative data collection and coding methods to test the framework. We solicited student participants from around the country; most were enrolled in an introductory (first-year) physical geology or Earth science class at our collaborating institutions. The diverse student population included geology majors, students fulfilling their general education science requirement, community college students, students at large public universities, and students from private liberal arts colleges with varying socioeconomic backgrounds.

Participants in this study ($n = 260$) volunteered anonymous written feedback on their experiences with the GCX apps. We analyzed the qualitative data contained within the student comments, identifying key themes such as *collaboration* and *learning styles*. Student comments were then compared to the identified themes until patterns emerged, as delineated by Stake (1995).

We used the three qualities of the m-learning framework for our analysis: *personalization, collaboration*, and *authenticity*. Through an examination of the data, we immediately saw evidence that these themes were present in the students' written feedback. We proceeded with identifying students' reflections on learning styles, repetition and reinforcement of in-class experiences, collaborative learning, and learning environments in relation to students' experiences with GCX.

Personalization: Learning Styles. A commonly cited benefit of GCX was that the experience appealed to multiple learning styles and participants benefited from content interaction. A student stated, "The best part of the lab activity was the ability to be able to listen to the reading from the narrator's voice explaining what was going on."

In addition to the auditory content delivery, numerous participants cited the benefit of both audio and visual engagement, stating, "We got to be active and I liked that it had a video and a voice. It helped me be able to visually point out faults and processes a little better. The pictures gave examples for us to be able to visualize the actual faults and processes." Another student shared, "By providing visual examples, along with detailed explanations though a hands on process, helped me grasp these concepts much better." The visual stimulation of GCX was integral in the students' experiences. Participants felt the photos used provided them with realistic representations of geological phenomena.

Many participants were aware of their learning styles and shared how they benefited from the heightened visual experience: "The lab activity helped me learn concepts taught in class because I am a visual/interactive learner." Students understood words and concepts more thoroughly when they had an image to compare it to, and the scenarios that accompanied each question enabled them to visualize the problem and come up with an answer.

Personalization: Repetition and Reinforcement. Students drew numerous comparisons between their experiences with GCX and their coursework. The activity helped students be more reflective of their learning in the course as a whole and identify areas of competency and weakness. For some students, GCX helped them identify gaps in their understanding of the coursework itself. Others indicated that it provided clarity and reinforced course concepts, with one participant stating, "I do believe that the lab activity helped [me] learn concepts taught in class because things that I was unsure that I didn't know, I actually did know. It made some things more clear as well." Several participants mentioned that GCX solidified classroom learning through the revisiting of concepts.

Collaboration. Working with groups and partners and sharing conversational spaces mediated by GCX on their mobile devices resonated with participants. The collaborative nature of the activities helped students network, work within their existing lab groups, and learn from their classmates. Students felt that the ARFT encouraged communication and cooperation. In order to maneuver successfully and expeditiously through the activity, participants found it beneficial to work together. A student shared, "My favorite part was getting to experience it with my classmates and working together."

Students competed to achieve the best result possible. One student noted, "My favorite part of the activity was the 'rewards' for getting the most answers correct. Getting a jet-boat ride or helicopter lift out of the Grand Canyon can be fun and silly, or at least the whole make-believe aspect of it," while another student shared, "I liked trying to get the most points in order to get a helicopter ride."

Authenticity. Active and interactive learning techniques help students acquire knowledge, develop critical thinking skills, solve problems in a variety of situations, and think independently (Bailey, 2004; Ratto, Shapiro, Truong, & Griswold, 2003). Immersion in the AR setting of the activity in this study provided authenticity for the participants. The immersion of the student in the physical setting impacted them in numerous ways. One participant shared, "I liked the basic concept [of the lab activity]. Walking around and discovering landmarks with fitting questions really gave it an immersive feel that a simple quiz doesn't quite match." GCX helped students see themselves as geologists, not just students in a geology class, with students sharing, "Real life application of geology was fun" and "We were out in the field. I felt like a scientist." Students interacted with each other and their surroundings outside of a traditional lab environment. A student shared, "The best part for me was actually getting out of the classroom and interacting with all my peers. The whole virtual experience was great and we got to learn in a different way."

CONCLUSION

Often students enrolled in general education geoscience courses are intimidated. They frequently have a preconceived notion that science is hard and not enjoyable. The GCX apps offer a real, easily accessible, and technical way to eliminate intimidation.

We faced many challenges during the testing of GCX. Weather conditions ranged from extremely high temperatures (over one hundred degrees Fahrenheit) to extremely cold, with snow on the ground. From reading the comments, it became evident to us that regardless of the location or conditions, students enjoyed themselves. When technical issues arose with the devices, they were seen as problematic. Software glitches required students to reorient the device and restart the activity, but the students understood the limitations of the technology and worked through them. Often, these technical glitches added another opportunity for the students to collaborate and problem solve.

Small groups of students collaborated to find answers, while also competing to earn the top reward. Occasionally, students requested to do the activity a second time to improve their score. Students admitted that the activity was fun: "It was a very fun experience and a good time just exploring outside."

While it is not expected that augmented reality field trips or virtual field trips will replace the traditional field trip, it is apparent from student comments that these tools can provide an engaging and fun learning experience, as well as bring the wonders of our national parks into our students' field of view. Nanny (1990) said, "At any given point, we are stuck in one location and cannot instantly see things in faraway places. We are stuck with our human size and cannot shrink ourselves to see microorganisms, nor can we become giants and stretch into space for a more global point of view." Geoscience learners everywhere are constrained by their geography; therefore, the virtual or augmented reality field trip is valuable as a mechanism that has the potential to allow students to experience faraway places, shrink to the size of a silicon tetrahedron, or become giant to gain a global perspective.

Virtual and augmented reality field trips on mobile devices provide a pedagogically sound way to access student learning styles (kinesthetic, auditory, and visual) in a short period of time. Students are very familiar with the game-like aspects of these tools (interactive, reward based, competitive) and the mobile smart device platform. Survey comments indicate that students were engaged in learning geoscience through the GCX experience. Through the GCX applications students became involved in the activity in a personal way that made the material more relevant to them. The activity built collaboration and developed communication skills in different ways than traditional group work and lab activities. The GCX platform provides instructors with an easily accessible tool to tap into the motivational aspects of learning.

While these augmented reality field trips were developed with Grand Canyon as the setting, it would be ideal to expand the concept to as many of the national parks as possible. This would provide students, regardless of geographic location, with access to learn from the unique natural features that the parks provide. In addition to geologic concepts, we believe that AR and VFT content needs to be expanded to include biology, ecology, climate change, environmental issues, and anthropogenic impacts. Ultimately, there is potential to increase student enjoyment and engagement within the traditional geoscience classroom.

References

Anderson, M. (2015). Technology device ownership: 2015. Retrieved from http://www.pewinternet.org/files/2015/10/PI_2015-10-29_device-ownership_FINAL.pdf

Andresen, L., Boud, D., & Cohen, R. (1996). Experience-based learning. In G. Foley (Ed.), *Understanding adult education and training* (2nd ed., pp. 225–239). Sydney, Australia: Allen and Unwin.

Bailey, D. (2004). Active learning. Retrieved December 22, 2007, from http://coe.sdsu.edu/eet/Articles/activelearning/index.htm

Bandiera, O., Larcinese, V., & Rasul, I. (2010). The impact of class size on the performance of university students. Retrieved from https://voxeu.org/article/impact-class-size-performance-university-students

Beetham, H., & Sharpe, R. (2013). *Rethinking pedagogy for a digital age.* Oxford, United Kingdom: Routledge.

Bursztyn, N., Shelton, B., Walker, A., & Pederson, J. (2017). Increasing undergraduate interest to learn geoscience with GPS-based, augmented reality field trips on students' own smartphones. *GSA Today, 27*(5), 4–11.

Bursztyn, N., Walker, A., Shelton, B., & Pederson, J. (2017). Assessment of student learning using augmented reality Grand Canyon field trips for mobile smart-devices. *Geosphere, 13*(2), 1–9. doi:10.1130/GES01404.1

Cook, V. A., Phillips, D., & Holden, J. (2006). Geography fieldwork in a "risk society." *Area, 38,* 413–420.

Dahlstrom, E., & Bichsel, J. (2014). Study of students and information technology, 2014 [Research report]. Retrieved from http://www.educause.edu/library/resources/study-students-and-information-technology-2014

Deslauriers, L., Schelew, E., & Wieman, C. (2011). Improved learning in a large-enrollment physics class. *Science, 332*(6031), 862–864. doi:10.1126/science.1201783

Friess, D. A., Oliver, G. J., Quak, M. S., & Lau, A. Y. (2016). Incorporating "virtual" and "real world" field trips into introductory geography modules. *Journal of Geography in Higher Education, 34,* 1–19.

Fuller, I. C. (2006). What is the value of fieldwork? Answers from New Zealand using two contrasting undergraduate physical geography field trips. *New Zealand Geographer, 62,* 215–220.

Habraken, C. L. (1996). Perceptions of chemistry: Why is the common perception of chemistry, the most visual of sciences, so distorted? *Journal of Science Education and Technology, 5*(3), 193–201.

Hill, J. L., Nelson, A., France, D., & Woodland, W. (2012). Integrating podcast technology effectively into student learning: A reflexive examination. *Journal of Geography in Higher Education, 36,* 437–454.

Hovorka, A. J., & Wolf, P. A. (2009). Activating the classroom: Geographical fieldwork as pedagogical practice. *Journal of Geography in Higher Education, 33,* 89–102.

Jacobson, A. R., Militello, R., & Baveye, P. C. (2009). Development of computer-assisted virtual field trips to support multidisciplinary learning. *Computers & Education, 52,* 571–580.

Kastens, K. A., Manduca, C. A., Cervato, C., Frodeman, R., Goodwin, C., Liben, L. S., . . . Titus, S. (2009). How geoscientists think and learn. *Eos, 90*(31), 265–272.

Kay, R. H. (2012). Exploring the use of video podcasts in education: A comprehensive review of the literature. *Computers in Human Behavior, 28,* 820–831.

Kearney, M., Schuck, S., Burden, K., & Aubusson, P. (2012). Viewing mobile learning from a pedagogical perspective. *Research in Learning Technology, 20*(1), 1–17. doi:10.3402/rlt.v20i0.14406

Lee, K. (2012). Augmented reality in education and training. *TechTrends, 56*(2), 13–21. doi:10.1007/s11528-012-0559-3

Liarokapis, F., Mourkoussis, N., White, M., Darcy, J., Sifniotis, M., Petridis, P., . . . Lister, P. F. (2004). Web 3D and augmented reality to support engineering education. *World Transactions on Engineering and Technology Education, 3*(1), 11–14.

Mazur, E. (2009). Farewell, lecture? *Science,* 323, 50–51. doi:10.1126/science.1168927

McGreen, N., & Arnedillo Sánchez, I. (2005). Mapping challenge: A case study in the use of mobile phones in collaborative, contextual learning. In P. Isaias, C. Borg, P. Kommers, & P. Bonanno (Eds.), *Mobile learning 2005* (pp. 213–217). Qawra, Malta: International Association for Development of the Information Society Press.

Mogk, D. W., & Goodwin, C. (2012). Learning in the field: Synthesis of research on thinking and learning in the geosciences. *Geological Society of America Special Paper 486,* 131–164. doi:10.1130/2012.2486(24)

Nanny, M. (1990). Interactive images for education. In S. Ambron & K. Hopper (Eds.), *Interactive multimedia* (pp. 87–98). Redmond, WA: Microsoft Press.

National Park Service Organic Act, 39 Stat. 535 (1916).

Orion, N., & Hofstein, A. (1994). Factors that influence learning during a scientific field trip in a natural environment. *Journal of Research in Science Teaching, 31,* 1097–1119.

Ozdemir, G. (2010). Exploring visuospatial thinking in learning about mineralogy: Spatial orientation ability and spatial visualization ability. *International Journal of Science and Mathematics Education, 8*(4), 737–759.

Prince, M. (2004). Does active learning work? A review of the research. *Journal of Engineering Education, 93,* 223–231.

Pringle, J. K. (2013). Educational environmental geoscience e-gaming to provide stimulating and effective learning. *Planet, 27*(1), 21–28.

———. (2014). Educational egaming: The future for geoscience virtual learners? *Geology Today, 30*(4), 147–150.

Race, P. (1993). Never mind the teaching—feel the learning! *Quality Assurance in Education, 1,* 40–43.

Ratto, M., Shapiro, R. B., Truong, T. M., & Griswold, W. G. (2003). The ActiveClass project: Experiments in encouraging classroom participation. In B. Wasson, S. Ludvigsen, & U. Hoppe (Eds.), *Proceedings of computer supported collaborative learning 2003: Designing for change in networked learning environments* (pp. 477–486). Dordrecht, Netherlands: Springer.

Rigney, J. W., & Lutz, K. A. (1976). Effect of graphic analogies of concepts in chemistry on learning and attitudes. *Journal of Educational Psychology, 68*(3), 305.

Sharples, M., Taylor, J., & Vavoula, G. (2007). A theory of learning for the mobile age. In R. Andrews & C. Haythornthwaite (Eds.), *The SAGE handbook of e-learning research* (pp. 221–247). London, United Kingdom: Sage.

Spicer, J. I., & Stratford, J. (2001). Student perceptions of a virtual field trip to replace a real field trip. *Journal of Computer Assisted Learning, 17,* 345–354.

Stainfield, J., Fisher, P., Ford, B., & Solem, M. (2000). International virtual field trips: A new direction? *Journal of Geography in Higher Education, 24*(2), 255–262.

Stake, R. (1995). *The art of case research.* Newbury Park, CA: Sage.

Student Transfer Reform Act, SB 1440, State Senate (2010). Retrieved from http://www.leginfo.ca.gov/pub/09-10/bill/sen/sb_1401-1450/sb_1440_bill_20100929_chaptered.html

Stumpf, R. J., Douglass, J., & Dorn, R. I. (2008). Learning desert geomorphology virtually versus in the field. *Journal of Geography in Higher Education, 32,* 387–399.

Tal, R. T. (2001). Incorporating field trips as science learning environment enrichment: An interpretive study. *Learning Environments Research, 4,* 25–49.

Traxler, J. (2007). Current state of mobile learning. *International Review on Research in Open and Distance learning, 8*(2), 1–10.

Tuckey, H., Selvaratnam, M., & Bradley, J. (1991). Identification and rectification of student difficulties concerning three-dimensional structures, rotation, and reflection. *Journal of Chemical Education, 68*(6), 460.

Whitmeyer, S. J., & Mogk, D. W. (2013). Safety and liability issues related to field trips and field courses. *Eos, Transactions American Geophysical Union, 94*(40), 349–351.

Yuen, S., Yaoyuneyong, G., & Johnson, E. (2011). Augmented reality: An overview and five directions for AR in education. *Journal of Educational Technology Development and Exchange, 4*(1), 119–140.

Young park visitor with smile. Courtesy of the National Park Service.

Partnering for the Next Generation of Learners

Just as the parks are rapidly changing due to climate change, the demographics of our country are as well. However, different from climate change, this demographic shift is going to make us stronger, more diverse, and hopefully more resilient in the face of ecosystem and economic challenges. The next generation of learners, often dubbed "Generation Z," were born between the mid-1990s and the mid-2000s. This generation makes up about 25 percent of the US population; thus they are already larger than the baby boomers and the millennials. Additionally, this generation is likely to be the most racially diverse group of US citizens that our country has ever seen. This section is based on the premise that we need the next generation to see the value of parks, learning, and civic engagement. These chapters argue that the National Park Service (NPS) must create meaningful partnerships if they are to remain relevant for the next generation. From centennial campaigns to overnight camps to classroom visits—the only way to engage the future is to work together.

In 2016, the NPS's centennial was recognized in a yearlong celebration that included creative programs, events, and activities designed to engage new audiences. The primary goal was to "connect with and create the next generation of park visitors, supporters, and advocates." This one-hundred-year anniversary was a watershed moment for the NPS to innovate and deepen engagement with partner organizations and visitors. In her chapter, "Place-Based Learning Fosters Engagement and Opportunities for Innovative Partnerships," Newton, who was working with the

National Park Foundation (NPF) at the time, explains how partnerships are critical to reaching the next generation of citizens and park goers. Newton discusses the *Find Your Park* and *Every Kid in a Park* campaigns and the evolution of the agency's leadership, which facilitated high-level, open-ended direction in planning for the centennial.

Yandala, Wright, and Sánchez introduce us to a long-term partnership that has created long-lasting results. In their chapter, "A Partnership Model of Education at Cuyahoga Valley National Park," they describe how Cuyahoga Valley National Park in Ohio leveraged a key partnership to build award-winning environmental education curriculum. It began when park leadership explored the potential for an overnight residential learning program. The need to create curriculum, develop long-term relationships, manage food service and custodial operations, and connect with teachers and new audiences would require a nimble and knowledgeable partner. Together, the NPS and a nonprofit organization, the Conservancy for Cuyahoga Valley National Park, built the award-winning Cuyahoga Valley Environmental Education Center. Working together, they have offered education programs serving more than ten thousand youth from urban neighborhoods every year.

Heading west from Ohio we are introduced to another unique, place-based partnership in "Pura Vida Inspires Diversity and Engagement at Grand Teton National Park." In 2010, Grand Teton National Park (GTNP) formed a partnership with the Teton Science Schools (TSS) to create the Pura Vida program. In this chapter, Freedman explains how this program engages local Latinx youth in GTNP and encourages participants to consider careers with the NPS. Almost one-third of the total population in Jackson, Wyoming, is Latinx and it is considered a GTNP gateway community, making this a significant program. The NPS understands that if public land is to remain relevant to future generations, then the agency must create initiatives like Pura Vida to engage diverse audiences.

Mounting research connects place-based, experiential learning with positive student outcomes, such as improved academic achievement and preparedness for life after high school. In addition, this model of education has been found to energize teachers and build relationships between schools and communities (Lieberman, 2013). In the chapter "What Really 'Matters' at Stephen T. Mather Building Arts and Craftsmanship High School," Shanley and Adams-Rodgers tell the story of a New York City high school that is thriving, innovative, creative, and energized by both the work and the students served. The authors detail the collaboration and creativity that built the NPS high school in the heart of Hell's Kitchen, New York City: the Mather High School. Learning from NPS experts, students study landscape management, historic preservation, masonry, decorative arts, and carpentry—bringing historic trades and craftsmanship into the current century.

The last chapter in this section, "Learning Historic Places with Diverse Populations: Making the Case for Teacher-Ranger Professional Development," illustrates another variation of collaboration and successful partnership building. View and Azevedo present an in-depth, qualitative case study, which chronicles a yearlong collaboration between NPS rangers and middle school teachers in Washington, DC. The rangers and teachers worked together to create an interactive, place-based curriculum that spanned more than a semester. This curriculum was designed to engage students in the pro-democracy movement by exploring significant locations and lessons from our social justice heritage. Thus, building on strengths of the instructional team members—NPS staff, middle school teachers, and university partners—they wove historical concepts and contemporary issues into student-led experiences at a diverse set of national historic sites in the Washington, DC, area.

Youth climbing at Obed Wild and Scenic River (US National Park Service). Courtesy of Dawn Kish Photography.

13

Place-Based Learning Fosters Engagement and Opportunities for Innovative Partnerships

SUSAN NEWTON

INTRODUCTION

In 2016, the National Park Service (NPS) Centennial was recognized in a yearlong celebration that included creative programs, events, and activities designed to engage new audiences. The NPS defined its intention to address the trend with its centennial goal: "Connect with and create the next generation of park visitors, supporters, and advocates" (NPS, 2016). Such a high-level, open-ended direction allowed for both creativity and innovation in programming, recognizing the unique opportunities and challenges of each location. This centennial goal created opportunities to examine the continuum of engagement with visitors as well as relationships with partner organizations. Prior to the centennial, the NPS outlined this shift toward developing innovative partnerships in several key strategy documents, including *A Call to Action*, *Revisiting Leopold*, and *Achieving Relevance in Our Second Century*.

A Call to Action

A Call to Action (NPS, 2012) outlined the NPS agenda for engagement as the agency entered the next century of service. This document made a clear distinction between the first century of the NPS and the second. The NPS's first century focused on stewardship and enjoyment of specific places; its second has broadened its reach into urban centers, across rural landscapes, deep within oceans, and across night skies. Its specific goals and measurable actions include

- connecting people to parks,
- advancing the NPS education mission,
- preserving America's special places, and
- enhancing professional and organizational excellence.

Revisiting Leopold: Resource Stewardship in the National Parks

Revisiting Leopold, developed by the National Park System Advisory Board Science Committee (2012), sought to examine the NPS's resource management policies first outlined in 1963 as the *Leopold Report* (officially *Wildlife Management in the National Parks*; Leopold, Cain, Cottam, & Kimball, 1963). The scope included all NPS natural and cultural resources. The report emphasized that ever-changing conditions required new strategies and approaches:

> Environmental changes confronting the National Park System are widespread, complex, accelerating, and volatile. These include biodiversity loss, climate change, habitat fragmentation, land use change, groundwater removal, invasive species, overdevelopment, and air, noise, and light pollution. . . . Cultural and socioeconomic changes confronting the National Park Service are difficult to overstate. These include an increasingly diversified, urbanized, and aging population, a transforming US economy, and constrained public funding for parks. . . . Simultaneously, scientific understanding of natural and cultural resources has dramatically expanded, continues to grow at an accelerating pace, and is becoming more quantitative and technologically sophisticated. (National Park System Advisory Board Science Committee, 2012, pp. 4–5)

Achieving Relevance in Our Second Century

Achieving Relevance in Our Second Century, developed as a part of *A Call to Action*, outlined actionable goals and strategies for interpretation, education, and volunteer employees of the NPS. It was designed to be interdisciplinary in nature and adaptable to employees across the system. In Julia Washburn's (the first Associate Director of Interpretation, Education, and Volunteers) opening letter, she articulated the importance of partners:

> The problems we face today are too complex to be solved in isolation. They require systems-thinking and the input of many good minds. . . . Indeed, it is

through collaboration and working in concert with our partners that our generation will succeed in carrying out the National Park Service mission. (National Council for Interpretation, Volunteers, and Education, 2014, p. iii)

With these initiatives so documented, the stage was set for a refreshed conversation about how parks and their partners can work together to connect with visitors and build a continuum of engagement opportunities. As the congressionally chartered national charitable partner of the National Park Service, the National Park Foundation (NPF) welcomed and embraced this call. The NPF works with partners ranging from groups comprised entirely of volunteers, to organizations with hundreds of professional staff, sophisticated facilities, and strong fundraising to support programming. With the breadth of this network, the NPF has a unique opportunity to support engagement of park goers, from "strollers to seniors." With this structure and charter, NPF and NPS teams work closely to identify key opportunities for NPF involvement and partnership innovation in the NPS's second century.

This chapter introduces a continuum of engagement that the NPF has implemented to frame our strategic intentions to engage park visitors. The continuum is a reference tool to help practitioners and partners create appropriate content for their communities and target audiences. Specifically, this tool can be used to (a) identify the types and variety of engagement and outreach programs across the country, (b) help determine measurable outcomes associated with the programs, and (c) serve as a means for starting dialogue of where gaps—or perhaps an overload—of program types are occurring. In sum, this continuum reminds us to develop programming that is appropriate for the different levels of potential visitor involvement with the national parks.

THE CHALLENGE

As noted in *Revisiting Leopold* (National Park System Advisory Board Science Committee, 2012), one area of concern for the NPS has been visitation, both overall numbers of visitors and their demographics. The *Annual Summary Report*, which includes demographic data from 1904 to the most recent calendar year, indicates that the number of visitors remained relatively flat, even declining by approximately eight million, between 1986 and 2013 (NPS, 2017). At the same time, the changing national demographics were not reflected in visitation to national parks. The NPS's report on *Racial and Ethnic Diversity of National*

Park System Visitors and Non-Visitors 2008–2009 (Taylor, Grandjean, & Gramann, 2011) highlights that visitors "were disproportionately white and non-Hispanic" and that "visitation differences by race/ethnic group seem not to have changed much since the previous iteration of the NPS Comprehensive Survey in 2000" (p. v). Compounding this issue of nonrepresentative park visitors was another trend: millennials and their increased use of technology to connect with each other and the world. A report in *Environment & Energy News* summarized this sentiment, explaining that NPS is "facing a new challenge: the need to connect with millennials—a generation of potential visitors who are more comfortable in front of a screen than a sweeping vista" (Hiar, 2015, para. 4). NPS leadership was concerned that these trends indicated that national parks were losing relevance with the American people.

THE APPROACH

In considering how to help these varied audiences connect to and engage with public lands, the NPF realized the necessity of employing a variety of methods. People experience parks differently—some are intimately aware of parks and their meaning, while others have no familiarity. Participants range from the very young to the very old, and at different times in their lives they may engage in very different ways. For example, elementary students may be learning about content, graduate students may be immersed in scholarly programs or research, and older adults and families may be using parks as places of recreation or lifelong learning. We believe that the full range of park visitors and volunteers fall along a continuum of engagement, from the unaware to those who fully support and actively advocate for national parks. As such, this continuum represents the broad range of potential audiences and key actions of each audience type.

The upper continuum represents the progression of an NPS visitor from someone completely unaware of park operations and opportunities (far left) to someone so thoroughly and personally invested in the parks that they advocate for them (far right). Note that at each successive move to the right, as involvement increases, the audience size decreases. The lower continuum, connected to the upper via a network of arrows, represents the actions necessary to move a visitor to the next level of the model. The continuum can be entered at any phase, as the entry point depends on the visitor's current level of park involvement.

The continuum provides a way to identify groups of park visitors and their needs. We introduce each of these visitor types and strategies for

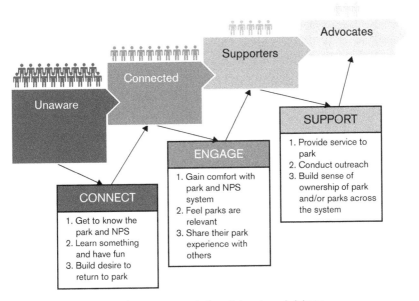

FIGURE 13.1. Continuum for engagement of participants and visitors.

engagement in meaningful park experiences, beginning with the largest audience, the *unaware*. This group is the most diverse, ranging in age, gender identification, race, and geography. Goals and strategies geared toward this audience require multifaceted approaches focused on introduction and awareness building about park resources and structures.

The next visitor type is the *connected*, which includes those who may be aware but may not feel NPS programs and services are personally relevant. Lack of relevancy can manifest in ways that include a lack of personal interest or appeal, limited perceptions about what park experiences entail, or lack of understanding of proximity to parks and programs. The disconnect can also be symptomatic of deeper issues, such as lack of access due to cost or transportation issues, cultural barriers, or not feeling welcome in national parks. The activities at this level of engagement should be designed for ease of entry, with few to no barriers for participation. Examples might include a daylong festival celebrating a new season, an invitation to view a night sky, or a guided hike. In general, these encounters are designed to be a one-time experience and are not dependent on a previous action for success. Although a visitor may engage with a park more than one time, the activities are not contingent on each other.

The *supporters* group includes those who are aware of national parks and have established a personal connection. National parks have relevance in these audience members' lives and they are motivated to engage and support the national parks. Goals and strategies in approaching this audience can become more complex in terms of level of activity, time frame of activity, repetitiveness of the activity, and dependency on previous activities. Examples include volunteering in parks, participating in citizen science programs, working on trail crews, or supporting a fundraising campaign.

The far right side of the continuum includes the park *advocates*, representing the highest level of engagement. Not only do these visitors support parks on a regular basis, they also turn their personal involvement into actions that promote the value of the parks for others. There is a wide range of activities that could be used for these advocates, but it should be noted that there are fewer people who pursue such deep engagement. For instance, an advocate could be a teacher who shapes curriculum to use parks as places of learning, an individual who promotes philanthropy for the parks by serving on the board of a local "friends of" group, or an individual who raises issues and concerns with their elected congressional representatives.

The continuum for engagement provides context for facilitating involvement; the NPS and other program providers can use this continuum to develop activities appropriate for the audience, topic, and place. For example, the strategic choices used to design the *Find Your Park/ Encuentra Tu Parque* campaign and the *Every Kid in a Park* (EKiP) programming illustrate applications of insights gleaned from this continuum.

FIND YOUR PARK/ENCUENTRA TU PARQUE

Find Your Park/Encuentra Tu Parque was a public engagement campaign focused on reaching new audiences, particularly multicultural millennials, and inviting them to explore the more than four hundred national parks across the country. The target audience represents the *unaware*, or nonvisitor, end of the continuum. Thus, the programming goal for this audience was to foster relevancy, connection, and support to bring new visitors to parks. The NPS and the NPF envisioned a far-reaching campaign, which evolved over a year and a half. The campaign required significant time and support from the NPS, the NPF, and a professional advertising agency. This campaign relied on a multipronged approach to reach audiences, including advertising, creating events and experiences, fostering partnerships, and promoting professional public relations activities. For instance, realizing that nonvisitors have difficulty seeing themselves in national parks, advertising played upon imagery showing how people and parks interconnect.

As the campaign progressed, opportunities were introduced to facilitate visitors' progress along the continuum, moving into the *connected* and *supporters* categories. While the NPF did not track individual progress from awareness to support, the arc of the campaign was structured to introduce and increase levels of engagement over time.

Throughout the campaign, NPF identified and evaluated measures of success for each of the goals along the continuum, which included increasing (a) relevancy (connection), (b) engagement, and (c) support. To measure relevancy, NPF monitored audience perceptions and understanding of the NPS (parks and programs) as shared on social media sites and through donated media time for campaign advertising. NPF measured engagement as increased and deepened public connection, estimated through social media engagements, visitation increases, uses of hashtags such as #FindYourPark and #EncuentraTuParque, and the survey-reported likelihood of millennials to visit a park. We quantified support by the number of new members in the parks community (e.g., followers through social channels and subscriptions to park-related communications) and indications of millennials' likeliness to volunteer (captured via surveys of millennials).

EVERY KID IN A PARK

The NPF has supported field trips to parks for K–12 students for several years. The goal of the *Every Kid in a Park* (EKiP) program is to help school children visit a nearby national park, build interest and confidence, and consider future visits. Within our continuum, EKiP programs can be designated as efforts to move students from the level of unaware to connected. Between 2015 and 2018, the NPF provided more than three hundred field trip grants to parks and their partners across the country, enabling 390,000 students to visit parks. Each site provides programming using a standardized three-part model:

- Target an unaware audience, including schools, teachers, and students who have not previously been to the park. This requires the park to find new partners and schools to work with, staying focused on one grade to reach new students each year.

- Increase opportunities for moving from unaware to connected, implementing a "three-touch" design that includes a presite activity, a field trip day, and a postvisit element.
- Get youth outside by transporting kids to parks (e.g., by bus, boat, train, public transportation, bikes, or walking).

At Obed Wild and Scenic River, more than 150 students from the Boys & Girls club in Knoxville, Tennessee (an hour from the park), visited the park for the first time. The combination of curriculum-based activities and challenging recreational experiences provided an inspiring introduction to the park.

Applied Research Northwest, an NPF partner, reported that feedback from the event was very positive, with one ranger calling it the best experience of their professional career, noting how it enabled the park "to reach out to new audiences and to youth who may of never had the opportunity to experience their national park" (Applied Research Northwest, 2016, p. 11). Upon completion of the program, the park leveraged students' and rangers' photos and experiences to help promote the park through blog posts, press releases, emails, and newsletters, helping the children build deeper connections and show support for the park (Applied Research Northwest, 2016).

To assess NPF's success in building connections between kids and parks, we surveyed the student participants regarding their enjoyment, their previous visits and familiarity with parks, and their interest in returning to the site. The survey results indicated that the students ($n = 1,661$) felt connected to the site; 63 percent agreed that they "definitely want to return" to the park.

RECOMMENDATIONS

Campaigns of the scope and scale introduced in this chapter may not immediately seem applicable to local parks and partners. However, these engagement strategies can be applied at all levels. For example, developing partnerships with local community influencers to share park experiences through their own channels, creating an ambassador program within the community, or using donated media through a local newspaper, radio station, or bus system are possible scaled strategies. With appropriate planning, *Every Kid in a Park* is an example of a scalable type of programming that can succeed with all national parks and their partners.

With the scalability of these programs in mind, here's a list of recommendations for building innovative partnerships to foster engagement in place-based learning:

- **Know your audience.** NPF targeted the *Find Your Park/ Encuentra Tu Parque* campaign at the unaware, nonvisitor, millennial audience. Research revealed that this audience relied on social media and web-based platforms for information. Additionally, the *Find Your Park* language was structured specifically as an invitation, rather than instruction, to reflect the target audience's communication style preferences.

- **Step outside your comfort zone.** Neither NPF nor NPS had ever launched a large-scale public engagement campaign. To achieve goals of reaching new audiences, they reached beyond "business as usual" practices and embraced innovative messaging and techniques.

- **Commit to multicultural sensitivity.** In the case of the EKiP event, organizers learned important lessons about overcoming communication barriers when a photo release form was not translated into Spanish. For *Find Your Park/Encuentra Tu Parque*, NPF continued to refine strategies to reach different audiences, taking an innovative perspective on communication platforms and outreach strategies.

- **Capture information to measure how well you achieve program goals.** These are key considerations for program design, planning, and implementation. It is critical to allocate appropriate time and resources at the onset of the program, to align program goals with measures of success, and to provide adequate time and funding for evaluation activities to take place. The insights allow for course corrections, improvements, or recognizing when the job is done.

- **Thank donors.** Measuring the program's success is also valuable for donor stewardship; it helps donors understand the impact of their philanthropic investment.

- **Recognize supporters.** Working in the national parks community requires supporters of all kinds—philanthropic, government agency, partner, industry expert, and more. Work with parks and partners to recognize supporters in the most appropriate way.

CONCLUSION

As the NPS enters its second century, a focus has emerged toward engaging the next generation of park visitors, supporters, and advocates. Concurrently, the NPS is reaching out to its partners in more explicit ways to work in a collaborative manner. As such, the NPF was able to take on new roles in cultivating the next generation of park stewards. Through work at national, regional, and local levels, the NPF drew on experiences of many partners and experts in the field to determine appropriate engagement strategies. That work, over the span of several years, illustrated the importance of clarifying campaign and programming goals, and forming a realistic assessment of where the target audience is in relation to those goals. These conversations and experiences produced a realization that a flexible engagement continuum can serve as a useful tool for setting appropriate strategies and accompanying metrics. With the continuum, local parks and partners can assess their current programming and outreach efforts. In turn, they can use the strategies and metrics to advance in areas that need improvement or further strengthen areas of success.

Acknowledgments

The author would like to thank NPF, the Washington Area Service Office (WASO), and all of the NPS staff involved in the development and execution of the *Find Your Park/Encuentra Tu Parque* and *Every Kid in a Park* campaigns. This research could not have taken place without the collective input from NPS employees, NPF employees, and NPF's advertising firm and evaluation team.

References

Applied Research Northwest. (2016). *Ticket to ride: 2015–2016 final report—Obed Wild and Scenic River*. Bellingham, WA: Author.

Hiar, C. (2015, April 16). National parks: NPS centennial aims to attract millennials, raise big money. Retrieved from https://www.eenews.net/stories/1060016905

Leopold, A. S., Cain, S. A., Cottam, C., & Kimball, T. L. (1963). Wildlife management in the national parks: Advisory board on wildlife management, appointed by Secretary of the Interior Udall. Retrieved from http://npshistory.com/publications/leopold_report.pdf

National Council for Interpretation, Volunteers, and Education. (2014, April 18). Achieving relevance in our second century: A five-year interdisciplinary strategy for interpretation, education, and volunteers as we enter the second century of the National Park Service. Retrieved from https://www.nps.gov /getinvolved/upload/IEVStrategicPlan_FINAL.pdf

National Park Service. (2012, August 24). A Call to Action. Retrieved from https://www.nps.gov/calltoaction/

———. (2016, December 19). National Park Service celebrates centennial success, looks ahead to building on momentum of 2016 [Press release]. Retrieved from https://www.nps.gov/orgs/1207/12–20–2016-centennial-success.htm

———. (2017). Annual summary report (1904 to last calendar year). Retrieved from https://irma.nps.gov/Stats/SSRSReports/National%20Reports/Annual% 20Summary%20Report%20(1904%20-%20Last%20Calendar%20Year)

National Park System Advisory Board Science Committee. (2012, August 25). Revisiting Leopold: Resource stewardship in the national parks. Retrieved from https://www.nps.gov/calltoaction/PDF/LeopoldReport_2012.pdf

Taylor, P. A., Grandjean, B. D., & Gramann, J. H. (2011). National Park Service comprehensive survey of the American public 2008–2009: Racial and ethnic diversity of National Park System visitors and non-visitors. Retrieved from https://mylearning.nps.gov/wp-content/uploads/2016/08/Racial-and-Ethnic-Diversity-of-National-Park-System-Visitors-and-Non-Visitors-2008-2009 .pdf

Students visiting Cuyahoga Valley National Park, Ohio. Courtesy of Conservancy for Cuyahoga Valley National Park.

14

A Partnership Model of Education at Cuyahoga Valley National Park

DEB YANDALA, KATIE WRIGHT, AND JESÚS SÁNCHEZ

For many in our country, when they hear the words *national park*, they think of high mountain ranges, wide open spaces, and miles of undisturbed wilderness. The areas that we have deemed to be important to the United States, and to the world—important enough to be managed by the National Park Service (NPS)—include magnificent natural wonders as well as sites that tell important American stories. Smaller parcels of land and historic places are significant for their location and their meaning for the past and future. Cuyahoga Valley National Park in Peninsula, Ohio, is one of the smaller places. These thirty-three thousand acres protect historical stories and valuable habitat adjacent to an urban area. From its enabling legislation and onward, the park has mandated and emphasized strong interpretation and education programs.

When park leadership determined that they wanted a residential learning program, one where students could stay overnight and have a multiday experience, the park superintendent and his staff did a careful analysis of the field of residential learning. They addressed the capacity of federal employees to create localized curriculum, develop long-term relationships with teachers, manage food service and custodial operations, and connect with new audiences. At the end of the assessment, the team concluded that leveraging the skills of a partner organization would provide the best opportunity for success.

Out of this conclusion came the award-winning and highly successful Cuyahoga Valley Environmental Education Center, comanaged by the

NPS and a nonprofit organization, the Conservancy for Cuyahoga Valley National Park. The park and this key partner offer high-quality education programs annually for over ten thousand children from local schools and communities. The Center's staff has, in turn, partnered with numerous schools and organizations to develop and deliver programs that are academically sound and engaging to students. The park is especially known for its work with urban school districts. As the future of national park education is developed, Cuyahoga Valley National Park provides a model of engaging with partnerships to provide innovative programs that engage young people in developing skills and values to shape the future.

CURRICULUM DEVELOPMENT

When the idea for a residential learning center was first proposed, NPS staff explored partnering with a local state university to develop and staff the educational program at Cuyahoga Valley National Park. When state budget cuts caused the university to withdraw from the proposed arrangement, the park turned to its primary partner (its *Friends Group*), the Conservancy for Cuyahoga Valley National Park, and asked if they would be the operating partner of the new Cuyahoga Valley Environmental Education Center. The group agreed. As a nonprofit, they were nimble and creative, and could easily hire staff with skills to run a residential center, create a curriculum, and begin programming.

From the beginning, the NPS and nonprofit staff agreed that engaging the community was critical to the success of the curriculum design process. They put aside two potential stumbling block notions: one is that simply getting students to a national park is adequate, and the other is that trained park interpreters have enough skills to develop programs on their own. Rather than assuming that the park would engage students on its own and that the rangers should develop all of the education from scratch, the team initiated a collaborative curriculum development process. They spent time with teachers, students, and community members to learn about their knowledge of the park, their interest in learning about the environment, and the educational methodologies that would best supplement classroom-based learning.

Teachers identified hard-to-teach and difficult-to-learn concepts that would be best taught in a field setting. Students expressed an interest in learning about the local natural environment. At the time, there was a trend to teach about tropical rain forests, yet students wanted to know more about what was in their neighborhoods and what they could do to protect these environments. Teachers and students knew that the Cuyahoga River captured the attention of people around the world when it "caught fire," yet they had little knowledge of the geography and biology of the river, or why it happened. This input prepared the NPS and Conservancy team for their dive into curriculum and the identity of the Cuyahoga Valley Environmental Education Center.

Using an educational advisory committee that included school district curriculum directors, administrators, and teachers allowed the nonprofit organization and park staff to provide input and feedback. Ready with this input, the Conservancy raised money to support curriculum development and contracted with area educators to assist. This process resulted in a curriculum that was aligned to state and school district curricular objectives. Further, it meant that area schools had immediate buy-in and were highly interested in the quality of the program.

A SHARED MODEL OF LEADERSHIP

As the Cuyahoga Valley Environmental Education Center was being developed, it was important to ensure it had strong leadership from the beginning. A leadership team was established that consisted of three NPS staff and three board members from the Conservancy, who determined the administrative structure for the Center. The Conservancy was given the overall administrative responsibility for the Center, with key input from park leaders. One important premise that remains to this day is this shared model of leadership.

The director of the Center is an employee of the Conservancy. The director works closely with the park's Chief of Interpretation, Education, and Visitor Services and the division's operations managers to make management decisions and establish future directions. Day-to-day responsibilities, including program design and teaching, are managed and implemented by both Conservancy and NPS staff, who work side by side. Today, an education committee of the Conservancy—which includes board members and community volunteers—provides guidance to the Center. The Center's director and NPS staff regularly engage with the committee to get their input and guidance.

When students participate in programs at the Center, they know they are in a national park. A uniformed park ranger participates in opening and closing activities; there are NPS arrowheads displayed in all the buildings; students receive Junior Ranger badges; and several park rangers teach programs. Within this model, program leadership is provided by Conservancy staff, and most of the Center's teachers work for the Conservancy.

The word most used by NPS and Conservancy staff to describe the partnership is *seamless*. It does not matter which organization staff work for—they have a common mission and have a shared culture at the Center. The NPS invests money in buildings and several staff positions, including one full-time staff member and several term or seasonal employees. Their investment results in over $1 million of staff time, program leadership, and program support for a high-quality environmental education program.

Most significant is the Conservancy's ability to raise money for the Center. The Conservancy has a sophisticated fundraising program and raises hundreds of thousands of dollars each year to assure that children from lower income families can participate in the Center's programs. The result is that the Center has a rich relationship with urban school districts in the area and reaches many children that would not have been able to attend without this financial model.

Because of the successful partnership model with shared leadership created at Cuyahoga Valley Environmental Education Center, there is an emphasis on extending the model to other partners. The park and the Conservancy both utilize partners for mutual benefit, from area educational organizations to youth-serving nonprofits. Over the years partners have helped the center serve a greater diversity of people and have carried the park mission into a wide variety of settings and populations.

LOOKING TO THE FUTURE

If our national parks are to be important to future generations of learners, then we need to help them see the relevancy of the park to the natural and cultural communities where they live. This is more than providing students a bus ride and a positive experience in the park. They need to have an opportunity to develop a personal connection with the park and to see how it is connected to the importance of their own community.

The staff of Cuyahoga Valley Environmental Education Center has embraced a mutual learning approach and is making connections in the communities that surround the schools being served. While it is helpful to partner with teachers who bring their students to the Center, the staff is deepening direct partnerships with communities so that the park is not such a different and unknown resource, especially to urban families.

This takes trust, relationship development, and an investment of time to build strong connections.

The Center is also exploring ways to enable students to see parks and environmental nonprofit organizations as opportunities for career choices. New programs are being developed at the high school level tied to the creation of career academies in local high schools. Some academies in urban areas are being developed around environmental themes and the Center is a key partner in their planning and implementation. The benefit to our national parks and their partners is the development of career pathways that might increase cultural diversity in environmental professions.

CONCLUSION

Cuyahoga Valley National Park created a larger, more extensive educational program by choosing a nonprofit to comanage and lead their education program. In turn, the partner has reached into surrounding communities to bring diverse partners to the table to strengthen the quality and reach of its learning opportunities. For educators, parents, and others who see value in connecting with our national parks as learning centers, engaging with park partners and parks can lead to educational success. Park partners strive to engage communities in national parks, and park-based education assures that we will have future generations of children committed to their national parks and enthused about preserving nature, history, and the values inherent in a national park movement.

Students gathered at Grand Teton National Park (Wyoming). Courtesy of Grand Teton National Park Foundation/Sheets Studio.

15

Pura Vida Inspires Diversity and Engagement at Grand Teton National Park

TEDDI (HOFMANN) FREEDMAN

Bella showed up to the first day of the program filled with anxiety, but those nerves were quickly washed away when she was greeted by a young Spanish-speaking woman with a friendly smile. This petite, energetic woman was wearing a broad-brimmed hat and green uniform. Bella quickly came to find out that she was a park ranger and the program leader. She was surprised to see someone who not only looked and spoke just like her, but who was also a park ranger in Grand Teton National Park (GTNP), a place that Bella had never visited despite growing up just twenty minutes away. From her very first interaction with this ranger, Bella knew this program was going to be different. She felt a sense of belonging and trust; she was excited to make new friends and explore new places. What she did not know, however, were the lifelong skills she would gain and the deep and personal connection she would make with the natural world.

In 2010, GTNP formed a partnership with the Teton Science Schools (TSS) to create a program called Pura Vida. Pura Vida seeks to better engage Latinx youth in GTNP and encourage participants to consider careers with the National Park Service (NPS). Almost one-third of the total population in Jackson, Wyoming, is Latinx. Jackson is also considered a GTNP gateway community, making this a significant group to engage in learning and discovery in the park. The NPS understands that if public land is to remain relevant to future generations, then the agency must create initiatives like Pura Vida to engage diverse audiences.

Place-based education serves as pedagogy for Pura Vida and provides a framework to enhance participant engagement in the park by drawing connections between the natural world and culture. The week-long program is free of charge, is nonresidential, and serves Latinx middle school and high school youth. Activities include hiking, canoeing, community service, guest lectures from GTNP staff, and a family dinner and celebration. Participants are encouraged to take ownership of their learning, identify connections between their culture and the local ecosystem, and apply leadership skills to be effective stewards.

Research suggests that Pura Vida can influence participant behavior and encourage increased engagement with the natural world. Even though Bella had never spent time in GTNP prior to Pura Vida, after just one week of exploring new places, she felt confident enough to bring her family and friends to the newly discovered spots. Her friend Yolanda, who was also a participant, felt similarly as she stated, "Pura Vida exposed me to all the great things GTNP has to offer. I started getting involved in other similar programs, which also helped expose me to the environment but also gave me leadership skills" (Hofmann, 2015, p. 37). She mentioned that the program enhanced her communication skills through participation in activities that required teamwork. Other participants echoed this sentiment, explaining that Pura Vida helped them to become familiar with GTNP. They were excited to share information with their families and friends about where to go for day hikes or picnics, and how and where to view wildlife from a safe distance.

The program's pedagogical focus on place also has the capacity to inspire participants to identify connections between the productivity of natural ecosystems and health of social communities. This encourages participants to further their learning and understanding of a place through exploration. Bella noted that

> staying and camping and actually learning to stay there and how to take responsibility and take care, I think that's really helpful, and it makes you be like "oh mom, I wanna go there!" And, like, tell your parents to do it. (Hofmann, 2015, p. 38)

Participants also became more concerned about the health of the environment and expressed the importance of respecting outdoor trails and spaces so that future generations could have similar experiences in nature.

Pura Vida places high value on the inclusion of family and invites parents, siblings, and relatives to participate in an evening campout as the program's culminating event. Participants and their parents enjoy the opportunity to engage in the outdoors together, as Marco explained,

> I love the inclusiveness of bringing your families into GTNP. That's a big benefit because not many programs do that. . . . Parents enjoy seeing accomplishments and hearing about their kids and their experience. . . . [It] gives a chance for parents not to be home and to be active. (Hofmann, 2015, p. 42)

One week after participating in the program, Marco and his family made plans to go camping together again in the same spot as they had the week prior during the program because it had been so much fun. Marco mentioned that his family was excited for the opportunity to get out and explore and that they felt confident in the camping skills he had gained during the program. Yolanda noted, "Pura Vida gets Latino families outside in nature. . . . Trying to bring more families and diversity into the park has been something that I've really loved and would like to continue to do" (Hofmann, 2015, p. 42). The program intentionally connects with families as a way to build a sense of community and belonging in the park.

The NPS has the opportunity to strengthen communities by offering more targeted programs like Pura Vida for people of all ages and backgrounds. Well-structured place-based education programs have the potential to foster a sense of empowerment and agency among individuals to engage in a place or community by supporting learning that is relevant and tangible. Pura Vida serves as an example of a program that has successfully increased the engagement of an underrepresented population in the national parks. The program continues to be led by park rangers who not only reflect the racial and cultural background of the participants but also can speak Spanish, which eliminates any language barriers and promotes a sense of trust and transparency. In addition, Pura Vida is free of charge and provides round-trip transportation, making the program accessible for many local families.

The program continues to grow in popularity from year to year. However, there are some remaining challenges. Initially, this program was thought of as a "pipeline" to help with diversity of NPS workers; however, for participants who qualify as Deferred Action Childhood Arrivals (DACA), their legal status prevents them from seeking permanent positions in the NPS. This remains a serious barrier for youth, and the NPS needs to find a solution to this issue to effectively recruit diverse populations into the workforce. Research (e.g., Hofmann, 2015) may serve as some of the groundwork for better understanding how place-based education programs can be used as a tool for encouraging engagement of different community groups in national parks.

Reference

Hofmann (Freedman), T. (2015). *Exploring the impact of Pura Vida on participant environmental behavior and engagement in Grand Teton National Park.* Science and Math Teaching Center Plan B Papers (Paper 38), University of Wyoming.

Carpentry student at Mather High School works on a historic preservation project. Courtesy of Mather High School.

16

What Really "Matters" at Stephen T. Mather Building Arts and Craftsmanship High School

DEBORAH SHANLEY AND LOIS ADAMS-RODGERS

INTRODUCTION

To tell the story of a New York City (NYC) high school that is thriving, innovative, creative, and energized by both the work and the students served, we must first recognize the journey. It began with the passage of the Organic Act, creating the National Park Service (NPS) on August 25, 1916, and the appointment of Stephen T. Mather as the initial director in 1917. The Secretary of the Interior, Franklin K. Lane, recognized that Mather possessed all the necessary leadership attributes: "passion, vision, excellent connections, money and a gift for public relations" (Heacox, 2016, p. 108).

Nearly one hundred years later in 2008, the journey continued with the newly formed 21st Century Commission, which was "charged with developing a 21st century vision for the NPS and for the magnificent collection of unique places it holds in trust for the American people" (National Parks Second Century Commission, 2009, p. 4). It was clear to the commissioners that this vision needed to include an expanded role of education (Chen & Jarvis, 2016).

FROM INCEPTION TO STRUCTURE

Mounting research connects place-based, experiential learning with positive student outcomes, such as improved academic achievement and preparedness for life after high school. In addition, this model of education has been found to energize teachers and build relationships between schools and communities (Lieberman, 2013). As our nation approached the NPS centennial, Chen and Jarvis (2016) reminded us of the important role of outdoor classrooms in connecting school and life lessons. In response, the Mather High School team embraced the idea of engaging in new types of learning in partnership with others.

Advocacy from Secretary of the Interior Ken Salazar and NYC Mayor Michael Bloomberg led to an official commitment in October 2012 to explore the creation of a new Career and Technical Education (CTE) high school. The proposed school would emphasize values of craftsmanship and the National Park Service ethic—to preserve cultural and natural resources for the enjoyment and use of future generations.

Key visionaries from different organizations (i.e., the director of Historic Architecture, Conservation, Engineering Center for the NPS Northeast Region; the commissioner of the National Parks of New York Harbor; and the superintendent of Governors Island National Monument) brought their lifelong passion to this project. The core team consisted of six NPS professional staff and one NYC Department of Education member, who were tasked with developing the curriculum, associated field and work-based learning experiences, professional development, and the required NY state CTE assessment. Together, they negotiated a shared vision for the Stephen T. Mather Building Arts and Craftsmanship High School, which opened with eighty-four freshmen in September 2013.

The journey continues in the heart of Hell's Kitchen, NYC, where this ninth grade to twelfth grade public high school caters to students who want hands-on learning experiences. At Mather High School, students could choose from fields such as landscape management, historic preservation, masonry, decorative arts, and carpentry. In 2019, Mather High School is a diverse learning community of students, with over 93 percent identified as students of color and 29 percent of students on individualized education plans (IEPs).

The commissioner of National Parks of New York Harbor, Joshua Jacobs, committed a position to this effort to ensure continuity and provide NPS technical guidance. A small NPS team advised the program's design, from the trades curriculum to alignment with the ideals and ethics of craftsmanship. Jacobs also led the effort to design units, lessons, and field experiences for the carpentry and masonry components, including the hands-on project-based trades instruction for the freshman class. He noted that he "became inspired to pursue the possibility of a full transition from practitioner to educator . . . [and] got an initial license to teach and was geared up to be a full-time teacher for year three" (personal communication, July 14, 2017).

Fortuitously, Jacobs successfully transitioned to being the full-time NPS staffer managing the partnership between the NPS and the NYC Department of Education at Mather High School. This illustrates commitment and alignment, which is critical to sustaining strong partnerships. In a complex partnership, such as this one, Jacobs noted,

> [in this] amazing job, I get to wear the hat[s] of preservation practitioner, educator and advocate while supporting Mather teachers' efforts to forge connections to the NPS. . . . [I provide] teacher professional development . . . in class instruction, [coordinate] hands-on project based field trips to NPS sites . . . [lead] Mather youth on spring break and summer preservation crews, [build] and strengthen partnerships . . . and [work] with the Mather HS advisory board.

This reflection demonstrated how Jacobs and other NPS employees work at Mather High School, going beyond teaching to include collaborating with content area teachers, engaging with the community, and beyond.

FROM SCHOOL DAYS TO STUDENT OUTCOMES

Imagine attending a career-tech class in a carpentry or masonry lab with no more than fifteen students. This allows for individual attention and balances the academic, social, and emotional components of this rich learning environment. It would not be unusual to have a geometry or trigonometry teacher teaching alongside a mason from Castle Clinton National Monument, a circular sandstone fort located in Battery Park in Manhattan. In addition, during a six-week unit in professional decorative plastering, there might be a full-time NPS staff person codeveloping the curriculum with content area teachers. A range of place-based internships could be available during the summer and school year, aligned with the mandated New York State Education Department curriculum. Several examples of this include

- working on a crew protecting and improving Wave Hill's woodlands, learning about restorative ecology, and enjoying the outdoor setting while making new friends (Wave Hill Forest Project; see https://www.wavehill.org/education/forest-project/);

- working at the Nature Conservancy's NYC LEAF Program to develop conservation leadership skills through hands-on environmental stewardship experiences (LEAF program; see https://www.nature.org/about-us/careers/leaf/learn-more/index .htm);

- joining a garden development program for youth through hands-on experiences in urban farming, sustainability, and nutrition at Harlem Growth Youth (see http://www .harlemgrown.org); and

- participating in field research studies focusing on ecological restoration, while receiving mentorship from working scientists.

Integrated classrooms and place-based, hands-on internships are keystones to this innovative education model. There is evidence of this success in the classrooms and also in the recognized strengths of Mather High School. The NYC Department of Education (2015) identified six transformative elements that drive school improvement and prepare students to compete and engage in the twenty-first century: (a) rigorous instruction, (b) supportive environment, (c) collaborative teachers, (d) effective school leadership, (e) strong family community ties, and (f) trust. Mather was rated to be excellent or good on all six indicators, demonstrating particular strength in an excellent school leadership team, collaborative teachers, and joyful rigor of instruction.

A CULTURE OF COLLABORATION

Principal Gabbard saw the potential for the program following the initial conversations with the NPS team. He was not disappointed in spite of what many describe as the culture clashes that happen when large public institutions are brought together in this type of endeavor. The facilitated conversations during the planning process built support and deep commitment on both sides. This was affirmed during the New York State Education Department program approval process. The examiners were struck by the deep partnership between Mather High School and the NPS. Most notably, students reported feeling cared for and the institutions' collaboration had a high degree of trust.

On June 28, 2017, Mather High School held its first graduation. The core values that were the foundation of the four-year experience were evident in the remarks by the students, the teachers, and the principal.

Of the fifty-four graduates, forty-seven were going to college, three were joining the military, and three had accepted employment with the NPS.

As the partnership approaches the beginning of the 2019–2020 school year, everyone continues to work together to ensure the partnership is sustained. A key challenge for institutions involved in this type of work is to develop and implement strong feedback loops to guide and improve collective practices. They must include, but not be limited to, recruiting, hiring, and retaining the right people and providing support and meaningful professional development opportunities. There is also value in engaging student, alumni, and parent perspectives throughout the process.

CONCLUSION

The success of Mather High School's model is embedded in their core values: Be action oriented, responsible, conscientious, and aware; promote positive risk taking; and engage in collaboration. In addition, the realities of shared vision, deep passion, tireless commitment to the mission, sense of humor, and well-articulated goals must be supported by all stakeholders. As John Fahey, former President and Chief Executive Officer of the National Geographic Society, noted, "Our spirits need the experiences that the National Parks offer. They help us fully understand what it means to be human" (National Parks Second Century Commission, 2009, p. 6). At the Stephen T. Mather Building Arts and Craftsmanship High School, the journey continues in partnership with the NPS.

References

Chen, M., & Jarvis, J.B. (2016, August 31). 100 years old, our national parks are the best outdoor classrooms. Retrieved from www.edweek.org/go /commentary

Heacox, K. (2016). *The national parks: An illustrated history.* Washington, DC: National Geographic.

Lieberman, G.A. (2013). *Education and the environment: Creating standards based programs in schools and districts.* Cambridge, MA: Harvard Education Press.

National Parks Second Century Commission. (2009). *Advancing the national park idea: National Parks Second Century Commission report.* New York, NY: National Parks Conservation Association.

New York City Department of Education. (2015). Framework for great schools and school quality reports. Retrieved from http://schools.nyc.gov/AboutUs /schools/framework/default.htm

Organic Act of 1916, 39 Stat. 535 (1916).

Outside the White House. Courtesy of Jenice View.

Learning Historic Places with Diverse Populations

Making the Case for Teacher-Ranger Professional Development

JENICE L. VIEW AND PAULA CRISTINA AZEVEDO

INTRODUCTION

Generally speaking, a National Park Service (NPS) ranger welcomes whoever visits their site with a warm greeting, instructions about the next showing of the site video, or the twenty-minute ranger talk. If the patron is accompanied by a school-age child, they may offer a Junior Ranger booklet with the temptation of earning a Junior Ranger badge. The ranger works from an interpretive plan that tells the story of the site.

However, there are more hurdles for a classroom teacher interested in making a field trip to the site. The teacher needs to convince the administration that the trip is connected to the curriculum and will not disrupt standardized testing (Neumann, 2013). The teacher may contact the site in advance but makes logistical arrangements at the school level and creates curricular links that help students understand the significance of the site. At the end of the visit, the ranger says goodbye and turns their attention to the next set of visitors. After the fact, students may remember the weather of the outing and may retain elements of the historical content, as evidenced by a teacher-created assessment. If no harm is done to students or the historic site, the visit is considered a success.

The conscientious teacher and ranger seek better measures of success that include a deep student understanding of the historical meaning of the site and its original residents. Rangers might hope that the youthful visitors become advocates and stewards for preserving the historic site, while teachers might hope that a field visit brings alive classroom-based history instruction that connects students with contemporary issues. The desire for these kinds of outcomes were apparent when we offered joint professional development to teachers and rangers called *Learning Historic Places with Diverse Populations* (LHPDP); LHPDP, an ongoing multiphase action research effort based at George Mason University, was partially funded by an NPS grant.

THE PROJECT

Phase II of LHPDP focused on a yearlong professional development and collaboration fostered among two embedded university researchers, Dr. Jenice L. View and Dr. Paula Cristina Azevedo; three NPS rangers, who will be referred to as Jill, Sarah, and Jason (pseudonyms are used to ensure participants' anonymity); and Miranda (also a pseudonym), a fifth-grade teacher with twenty-four students. Miranda's students were diverse in terms of race, ethnicity, language and cognitive ability, and gender.

We sought to share our love of history and historic sites with fifth graders in a public elementary school in the Washington, DC, metropolitan area by challenging the belief that history is one of the most hated subjects in school (Loewen, 2007). Also, we wanted to close the gap between teachers and historic sites (Griffin, 2004; Orion, 1993; Parsons & Muhs, 1994; Sacco, 1999; Xanthoudaki, 1998). We were curious to learn how a yearlong professional development process and collaboration between a classroom teacher and interpreters would impact the adults' professional practice and student understanding of civil rights activist Mary McLeod Bethune. In this chapter we describe the team's collaboration process and outcomes, with recommendations for expanding this experience with NPS partners.

METHODOLOGY

Data sources included a literature review, ethnographic notes/memos, classroom and field trip observations, interviews with teachers and rangers, student focus groups, student writings, and individual student interviews a year after the fact. The methods used to address the gathering and analyzing of data included collaborative action research (Zeichner, 1993), inquiry regarding institutional practices (Elliott, 1991), ethnographic case study (White, Drew, & Hay, 2009), narrative interviews (Wengraf, 2013), and constructivist grounded theory (Charmaz, 2012).

Researchers also investigated how the three elements of action research informed the structure of LHPDP. They examined the improvement of practice: the professional practices of the teacher, rangers, and researchers; the *understanding* of practice, such as the professional practices of the teacher and rangers; and the situation in which the practice takes place, that is, the K–12 classrooms and cultural historic sites seeking better community outreach and connections with K–12 classrooms (Carr & Kemmis, 2003).

MAKING THE CASE FOR TEACHER-RANGER PROFESSIONAL DEVELOPMENT

Many barriers challenged the professional collaboration and implementation of the curriculum, instruction, and interpretation. These included the fifth-grade ancient world history curriculum, the 2013 federal government shutdown, eleven snow days, and the pressures of standardized testing on students and teachers. Despite these challenges, data suggested that when teachers and interpreters collaborate to design integrated educational activities—including interpreter previsits to the classroom, relevant in-class curriculum, site visits with interactive interpretation, and postvisits with interpreters—elementary students developed historical skills through field trips to historic sites. Essential to the outcomes were (a) site visits (as opposed to virtual tours), (b) teacher skills, (c) interpretation skills regarding the historical content, (d) interpreter comfort with children, and (e) interpreter experience working with teachers. Additionally, the facilitating presence of the university-based researchers for data collection, logistical support, and oversight helped to sustain the teacher-interpreter relationship. In the following sections we describe what the researchers and participants learned from this unique professional development collaboration. The key findings fit into four main categories: (a) professional development; (b) collaboration; (c) development of meaningful curriculum, instruction, and interpretation for K–12 students; and (d) the impact of cross-profession professional development and collaboration on students.

Professional Development

The plan was to identify a limited number of rangers and teachers and have four face-to-face and two webinar sessions on the following topics: diverse populations and culturally relevant pedagogy, history versus fairy tales, a historic site as a primary source, gender issues and history, assessments, and a site visit to the Bethune Council House. We formed the team and began in fall 2013, but due to unforeseen challenges, the planned discussion topics, learning activities, and working sessions had to be modified. As the team formed, Jenice and Paula's academic and professional development expertise became less important than their skill as project facilitators. Instead of developing content and professional skill building, they instead facilitated the relationships and the logistics while documenting every activity. Interview data suggested that the researchers were relevant to the process. Their constant presence, data collection, logistical support, and oversight were the glue that held it together as they provided infrastructure within the professional development that allowed Miranda, Sarah, and Jason to move past the barriers to do their best work. Miranda was particularly effusive about the role of facilitation in serving the larger collaboration:

> This experience, as a teacher and then . . . sharing it with my students, was a phenomenal opportunity to see them learn and grow about places that are in their backyard that they haven't had a chance to experience. . . . Being able to collaborate with colleagues, who are in the [Washington, DC] area, but of course not in [the county] public schools, was an outstanding experience.

There is no evidence that Miranda's teaching *improved* in any particular way due to the professional development, because we were not measuring this skill, but it was an essential factor in project outcomes. However, as Bowling (2013) suggested, networking and developing partnerships with the community supported interpreters' skills. This was evident when Jason acknowledged that Miranda was a master teacher, with skills that could inform his interpretation skills:

> Something that reinforced really well watching her in the classroom, I think one of her techniques to get people to listen . . . I can't remember the exact thing, but everybody clapped . . . "if you are listening, you clap once." My mind is always [asking] how can I use some of this stuff with . . . adults on programs as well . . . people like being a part of a group . . . you foster that group, togetherness, constantly really.

With a clearer vision of Miranda's students, Jason and Sarah considered the effect of having ten-year-olds visit their sites, in a larger-than-typical group. This group of students would be coming with an "agenda" to link their fifth-grade world history curriculum with two historic sites in the nation's capital. One of Sarah's ideas from the fall professional development helped highlight a difference between "generic kid activities" and "student activities." To avoid making the visit to Bethune Council House a "furniture tour," she proposed an interpretive activity in keeping with the historic usage of the site:

> I think there should be an activity. I think they'll get more out of it if they actually do something in the house, like what the NCNW [National Council of Negro Women] did. This is the house for doing, not sitting around. So I really like the idea of creating some sort of product. . . . I think it might be cool to make a banner to go in the conference room and say, "This is where the NCNW accomplished their important goals. This is where they conducted business." So we are going to accomplish something in this room, and we are going to make a banner that we're going to use to tell the president about some things that we want changed. We're going to stand up for ourselves, and we're going to do this work here in this room, and we're going to take it out to the public.

This idea sparked some excitement among the team, including Jason, who envisioned an active type of interpretation that fully used both spaces in historically accurate ways—the inside of the Bethune Council House as a hub for political strategy and activism and the outside of the White House as a site for the expression of several First Amendment rights. However, Miranda expressed strong concerns about drawing negative attention to her elementary school, the individual students, or her own teaching should there be an uncontrollable outburst from a student, unexpected media presence, or political scrutiny during a student protest in front of the White House. The clash between expected in-school and out-of-school behavior created an intriguing debate.

The professional development impacted the rangers differently due to prior professional experiences and knowledge. Jason, an undergraduate history major, was more comfortable with historical content than Sarah, and he demonstrated a great deal of historical skill. He also demonstrated more comfort with children, partly because of his personality, and because the White House has more ready-made, on-site materials and ideas for interpreting the site for children. His prior experience and knowledge was obvious during his interpretation of the White House during the field trip. His interpretation was nuanced at the outset and his skill allowed him to engage in a constructivist interpretation. For example, a religious advocate disrupted Jason's intended presentation and he was able to quickly reframe his presentation to make a larger point about democracy, protest, and the symbolic power of the White House.

Sarah had much more experience with teachers than with children, and the Council House had few kid-friendly materials, although countless on-site artifacts. She reported that her interpretation skills seemed to expand over the course of the project:

> [The ideal professional development for interpreters is] a program like Learning Historic Places, something where rangers work directly with teachers, where I'm getting education-based training outside the Park Service. I think working directly with educators is the best way to see what teachers need, what they have to deal with, the type of programs they're looking for. And this partnering with teachers is . . . the best way to learn how to do those because . . . I've never been a teacher.

Nevertheless, the story Sarah told during the field trip suggested that, even as a ten-year-old child, Dr. Bethune could "do no wrong," and there was a tendency to celebrate her as a perfect role model. Like many NPS sites, the Council House seemed to mandate a heroic interpretation, which begs for more historical and interpretative skill and a better understanding of the cognitive and analytical skills of ten-year-old students. We saw that with more collaborative professional development, Sarah could have learned more from Miranda about how (and what) children learn in order to expand her interpretive practice. Likewise, Miranda could have stretched her required curricular content to accommodate place-based instruction. Both could do more to avoid the hero worship of textbooks and the furniture tours of historic sites.

Collaboration

The intention of the cross-profession professional development was for Miranda, Sarah, and Jason to collaboratively create in-class and field experiences that would lead to student products that might potentially be featured at the Bethune Council House. However, due to aforementioned challenges, and a major required project for all district fifth graders, the idea of students creating a product for display in the Bethune Council House was unrealistic. These realities of working with K–12 students and teachers demonstrated the importance of collaboration that is flexible and that addresses the realities of each profession and expectations of each individual and the group as a whole.

Miranda remained mindful of in-class and out-of-class assignments that supported success for all of her students, particularly those who were English-language learners, those with special learning needs, and high achievers. This did not hinder the team's efforts to build an educational program around the two selected sites. Miranda communicated clearly to the cross-professional team the demographics of her class and how she differentiates lessons in order to meet all her students' needs. This helped the rangers understand and appreciate the daily instructional challenges Miranda faced, and also provided Jason and Sarah with an understanding of how to develop interpretive programs for her students. This was evident when Jason and Sarah conducted a pre–field trip visit to Miranda's class.

Individuals in the cross-professional collaborative team shared insights of working with each other and their appreciation of the others' work and input. Sarah shared, "I think that's because Miranda's so proactive and amazing at her job. . . . She's innovative. I think it takes innovative people at this point to make something like this work."

As Miranda began her spring semester planning, she emphasized the value of collaboration:

> Obviously I can create activities and get ideas, but your [referring to interpreters'] integration or your suggested ideas or things to do are also helpful to me as a part of this process. Whether those things take place at Bethune or President's Park, it will be immensely helpful to me as we team together to work out activities that really stimulate thinking and benefit the kids.

Curriculum, Instruction, and Interpretation

In November we held the first face-to-face session at the Bethune Council House. We used the building as "text" and "artifact" and considered ways we might engage students in learning about Dr. Bethune, the National Council of Negro Women, racial segregation, women's rights, and leadership. There seemed to be evidence that the team was beginning to cohere as we discussed common language and goals for the classroom instruction and on-site interpretation.

At the second November session, Miranda and Jason struggled over the central focus and student activity on the field trip. Such struggles are natural in any collaborative effort, but were perhaps more apparent in this case with two professionals looking at the same set of students with different perspectives. Where Jason saw an opportunity to have students create banners at the Bethune Council House—as did women prior to the 1963 March on Washington—and enact a protest or public demonstration in front of the White House around issues that had meaning for them, Miranda was concerned about student safety, the school's public profile, parental and administrative consent, and the pedagogical value of such a demonstration. Jason chafed a bit at Miranda's resistance concerning one of the proposed activity ideas on freedom of speech:

I love the plan that we came up with and we talked about. And . . . what frustrates me the most is I think these kids could totally handle protest and demonstrations. And you give me a couple minutes, give me some thinking time . . . I could sell protest with a line that these kids would really understand.

Nonetheless, each participant recognized and respected each other's professional expertise and what everyone brought to the team.

As previously mentioned, one of the challenges of this collaboration was the fifth-grade social studies curricula. Miranda was so well versed in the fifth-grade curricula that she artfully embedded the lessons she created about Bethune, Eleanor Roosevelt, the Council House, and the White House in both history and English language arts. For instance, in December, Jill visited the classroom for the first time, in her NPS ranger uniform. She shared a PowerPoint presentation that made seamless connections between the leadership of Confucius and Mary McLeod Bethune, and the political power contained in the Council House and the White House. Students were excited about holding Jill's ranger hat and the prospect of visiting the White House in the coming months. After six professional development workshops designed and led by the university researchers, the team identified a course of action, led by Miranda's outline.

During the rangers' classroom visit in January, Jason showed a photo of the White House and then passed around a piece of limestone from which the building is constructed while posing questions to students. Sarah reminded students of Jill's visit and showed a PowerPoint of natural and cultural-historical sites while posing questions about national parks. Students received bookmarks and other souvenirs from each of the sites, with indications of more to come.

To prepare for the trip, Miranda designed several writing and critical thinking in-class activities exploring the basic background information on Dr. Bethune, the relationship between Eleanor Roosevelt and Dr. Bethune, definitions of "heroes" and "leaders," using inference to understand how to interpret history and cooperative action to build power, and using historical information to make decisions about the future.

After a great deal of planning, it was finally time for the field trip. As the bus approached the Bethune Council House, students observed that the building looked like a "regular" house in a residential neighborhood; some were surprised by the size. Upon entry, Sarah and two of her colleagues greeted the students; they were guided to a row of seats in the dining room area to listen to a minilecture about Dr. Bethune and the house. They were divided into two groups to tour the house, pose questions, and take photos.

Jason joined the students at the Council House during lunch. He joked around with them and posed for photos outside the house before everyone boarded the bus for the White House. Outside the White House, Jason pointed out the building's various sections and their purposes, posed and entertained questions, and shared a photo of women's suffrage protesters from the early twentieth century. Students unfurled a banner that they had made the day before featuring images and words that were important to them regarding leadership. Jason later described his thoughts on his interpretation:

> Everything went . . . as well as it could on Pennsylvania Avenue when it's a busy day like that. I don't know if I'd really put it in a negative, that we had the guy talking about the Bible. That's just something you have to deal with on Pennsylvania Avenue. . . . Yeah, there were distractions, but it did tie into what we were trying to talk with the kids about, pressing . . . and protesting, and demanding your rights and such.

There was a final in-class ranger visit. During this time, students rotated through four stations that Miranda had set up, each with a laptop featuring the website of an NPS site. Jill, Sarah, and Jason occupied the stations to describe the White House, the Bethune Council House, and other NPS sites in more detail and answer student questions. At the station with the university researchers, students were asked what they remembered most about the field trip, Dr. Bethune, the White House, their banner, and history in general. At the end of the visit, the rangers distributed and stamped NPS Passports with Bethune and White House stamps.

Impact on Students

Student reflections suggested the retention of specific details related to the Bethune Council House and the White House, the field trip, and student interactions with Jill, Sarah, and Jason. The reflective writings also provided evidence of student historical thinking skills as related to the project activities.

Miranda perceived that project objectives were met and that students felt a personal connection with Dr. Bethune, the Bethune Council House, and the White House, but her evidence was not triangulated; therefore, no one could make the claim that the intervention of LHPDP created personal connections. However, a year after the field trip, students had fond memories of their experience meeting the rangers and going to the Council House and White House. One student even expressed deep historical thinking a year after the field trip:

> I would probably say that while I was in the Bethune House, I learned a lot more history than just civil rights and women's rights. I learned that—well, first, when you read the textbooks nowadays, there's no Mary McLeod Bethune in there. So what I learned about her, it was like, wow, there's someone else in there, in the Civil Rights Movement that is really important because without her, maybe the Civil Rights Movement would have lacked a little bit. (interview, June 12, 2015)

At the end of fifth grade, any evidence of student attachment to history was not as a result of *Learning Historic Places with Diverse Populations*, but instead due to already-growing personal interest. Researchers found a spectrum of student attitudes toward history: Ian (white) hated everything in school except sports; Morris (white) questioned the value of learning history; Rahman (Somalian) was only interested in the history of Islam; and Carl (Latino) was curious about many subjects, including history. Only Beth (Asian American) was completely enamored of Dr. Bethune. Comparing the student utterances across all themes (January vs. June 2014) revealed that there were twice as many regarding historical skill as regarding historical content. Our observations were that Miranda emphasized skill over content and the students mirrored that emphasis.

There was no evidence at the end of the school year to show explicit links between student learning and teacher or interpreter skill. This could be read as "no news is good news," since ten-year-olds might only express a strong opinion if the instruction or interpretation were poor. It is understood, however, that the researchers' understanding of the impact of interpretation on the study participants was less secure than assessing what happened with in-class activities. This was partly due to the constant evaluation that naturally happens in the classroom (homework, tests, projects, etc.).

A year later, interviews with twelve of the students in their new middle school suggested that teacher skill, ranger skill, and place-based learning may positively impact the retention of historical content and student attachment to history and historic sites, regardless of self-professed love of history as a subject. All of the students indicated that they enjoyed the field trip. Five students, all girls, indicated that history was their favorite subject. For one girl, middle school American history overtook math as her favorite subject. Six of the students said that history was "okay" as a subject; five liked it better in middle school. One student indicated that she liked history better at Queen Elementary school, because of the visits by the LHPDP rangers and the interviews with the researchers.

In examining the comments from two students, who were the outliers at either extreme, we see positive impact from the fifth-grade participation in LHPDP. Sophia, who loved history and regularly visited NPS sites, remembered a great deal of information about Bethune's work, life, and home. She explained,

> I really liked how we were able to go to different places, and you would come and talk to us about it. That really got me into history a lot, too, because in fourth grade, I did—I noticed that history like sort of had a spark in me, but it wasn't like there in a way. . . . I do still use the passport, too. (interview, June 5, 2015)

Rahman disliked history in elementary school because he preferred learning about his religion and its history. By middle school he had come to appreciate that "knowledge is power, I guess, so [I] have to learn so I can get educated" and remembered some of the details of the Bethune House and a Muslim person talking in Arabic out in front of the White House on the day of the trip. He enjoyed having the rangers come to the classroom after the field trip to go through the NPS learning stations, and he and his family occasionally visited the National Mall.

The project seemed to foster an appreciation for historic sites that could contribute to students' long-term stewardship of these community resources. Several students expressed their appreciation of the rangers visiting their fifth-grade class. One student explained a year later, "I actually liked last year because . . . a lot of people will come [sic] to our classroom and teach us stuff, and we'll go [sic] like to the places we were supposed to go." Another student shared similar sentiments:

> Well, I really liked how we were able to go to different places, and you would come and talk to us about it. That really got me into history a lot, too. . . . But I really liked how you guys came in, and then we got to go to field trips and things like that. . . . I look at [the NPS passport], and I still know where I want to go.

In summary, the findings suggest positive value in bringing together NPS interpreters from related sites, a master classroom teacher, and researchers/facilitators with a goal of having a diverse group of students learn and retain historical knowledge about a little-known NPS site.

RECOMMENDATIONS

Cross-Professional Development

A goal for creating interpreter-teacher collaborations is to have children engage in an explicit examination of a historic site and why it exists. We suggest that this kind of examination could elevate the importance of historic sites, beyond the excitement a field trip, as a way of nurturing historical thinking skills and content knowledge. These types of projects also present opportunities for cross-professional development.

Collaboration among teachers and interpreters prior to field trips could help to create cohesive learning experiences. In addition, NPS staff and teachers could develop tools for assessing impact (e.g., student surveys), as classroom connections are an efficient way to increase site patronage. Such tools could also help resolve a lack of feedback for rangers, common at NPS sites where their connections with visitors are typically very limited.

Another idea was proposed by Jason. He suggested initiating a Ranger-Teacher-Ranger program (the inverse of the popular NPS Teacher-Ranger-Teacher program) that would allow interpreters to spend significant time in a K–12 classroom. This could help them learn more about child and youth development, pedagogy, curriculum, and assessments, which in turn could deepen their practice of interpretation. Such an experience could be a low-cost part of an interpreter's professional development plan.

Programmatic Sustainability

Creating ongoing partnerships between NPS sites and K–12 schools takes sustained effort. Some suggestions to deal with this include transitioning between stakeholders and ever-present funding issues.

Critical to our work was including a specific transition period, which allowed for an exchange of information between outgoing and incoming partners. This included the archiving and sharing of interpretation plans, and contacts and their information. We found all of these to be essential to sustaining and maintaining partnerships.

Funding projects like this is often a challenge. However, we found that this type of collaboration seemed to be less dependent on financial resources than on mutual will. Other costs were spread among several entities or funded privately.

Facilitator

When challenges arose, we found that a facilitator could provide creative solutions to help resolve real or perceived challenges. A facilitator should be comfortable shifting between the role as point of contact and as content provider, questioner, listener, and observer throughout the collaborative process. In our experience, the facilitator ensured the physical space and time for professionals to meet and she also tended to administrative details, such as securing transportation. The facilitator was able to provide a consistently safe space for all team members to achieve their goals.

Communication

There is a need to maintain frequent, open, and honest communication in order to advance the collaborative process. Each team member must be comfortable leading and delegating leadership roles to the other team members. Among NPS interpreters, use of in-house social media sites to share ideas and communicate within sites could be a way to promote organic collaborative projects.

Joy

Tilden (2007) explained that "[Interpretation] is something individual, something that comes from knowledge and doing, but you've got to feel it" (p. 17). He was passionate about educating youth and encouraged interpreters to share their passion and joy with young people through love of the work and what it represents. In this way, the success of efforts such as these depends on the extent to which levity and joy emerge while interpreters work with youth populations.

CONCLUSION

The significance of this study relates largely to curriculum and instruction in history, social studies, and interpretation studies. In 1990, Vincent Harding argued,

> [A] misunderstanding of the Black freedom movement—and therefore of the history of this country—had dire consequences for everyone, particularly if it fails to identify the movement as a central point of grounding for our own [U.S.] pro-democracy movement. (p. 107)

In the United States, the modern Civil Rights Movement is taught in K–12 settings "as a spontaneous, emotional eruption of angry but saintly African Americans led by two or three inspired orators" (View, 2004, p. 3) who emerged between the years 1954 and 1968—perhaps to emerge again in Senator Barack Obama's 2008 campaign to become the first African American president: "Rosa sat, so King could march, so Obama can run, so our children can fly" (Mosley, 2008, para. 5). Yet, this quote contributes to the misunderstanding against which Harding warned. Changing the paradigm of history instruction should include (a) place-based learning, (b) a mission-driven rather than ego-driven orientation to learning about "heroes/sheroes," and (c) collaborative content design and delivery among interpreters and teachers. Such a shift would be a step toward engaging students in one important segment of the U.S. pro-democracy movement and recapturing significant components of that social justice heritage.

References

Bowling, K. A. (2013). Understanding implementation of key best practices in National Park Service education programs. *Journal of Interpretation Research, 18*(1), 83–86.

Carr, W., & Kemmis, S. (2003). *Becoming critical: Education, knowledge and action research.* London, England: Routledge.

Charmaz, K. (2012). The power and potential of grounded theory. *Medical Sociology Online, 6*(3), 1–15.

Elliott, J. (1991). *Action research for educational change.* Buckingham, England: Open University Press.

Griffin, J. (2004). Research on students and museums: Looking more closely at the students in school groups. *Science Education, 88*(Suppl. 1), S59–S70. doi:10.1002/sce.20018

Harding, V. (1990). *Hope and history: Why we must share the story of the movement.* Maryknoll, NY: Orbis Books.

Loewen, J. (2007). *Lies my teacher told me: Everything your American history textbook got wrong.* New York, NY: Touchstone Press.

Mosley, K. (2008, November 2). Vote for hope. *Pittsburgh Post-Gazette.* Retrieved from http://www.post-gazette.com/opinion/Op-Ed/2008/11/02/Vote-for-hope/stories/200811020185/

Neumann, J. W. (2013). Teaching to and beyond the test: The influence of mandated accountability testing in one social studies teacher's classroom. *Teacher College Record, 115*, 1–32.

Orion, N. (1993). A model for the development and implementation of field trips as an integral part of the science curriculum. *School Science and Mathematics, 93*(6), 325–331.

Parsons, C., & Muhs, K. (1994). Field trips and parent chaperones: A study of self-guided school groups at the Monterey Bay Aquarium. *Visitor Studies: Theory, Research and Practice, 7*(1), 57–61.

Sacco, J. C. (1999, May). *Crafting exhibit experiences for school audiences.* Paper presented at the American Association of Museums 94th Annual Meeting, Cleveland, OH.

Tilden, F. (2007). *Interpreting our heritage.* Chapel Hill: University of North Carolina Press.

View, J. L. (2004). Introduction. In D. Menkart, A. D. Murray, & J. L. View (Eds.), *Putting the movement back into civil rights teaching: A resource for classrooms and communities* (pp. 3–12). Washington, DC: Teaching for Change and Poverty and Race Research Action Council.

Wengraf, T. (2013). BNIM short guide bound with the BNIM detailed manual. Retrieved from http://www.case-stories.org/narrative-interviews-1/

White, J., Drew, S., & Hay, T. (2009). Ethnography versus case study: Positioning research and researchers.. *Qualitative Research Journal, 9*(1), 18–27.

Xanthoudaki, M. (1998). Is it always worth the trip? The contribution of museum and gallery educational programs to classroom art education. *Cambridge Journal of Education, 28*(2), 181–195.

Zeichner, K. M. (1993) Action research: Personal renewal and social reconstruction. *Educational Action Research, 1*(2), 199–219.

Ranger launches canoe with youth. Courtesy of the National Park Service.

Strategic Intention for Park Learning and Practice

Learning in parks has generated a surge of boundary-spanning research. Scholars and practitioners are working together to design, implement, and evaluate park-based learning initiatives at multiple scales. Whether we borrow from other fields or focus on field-based curriculum, there are dozens of innovative approaches to deepen our understanding of park learning and practice. This section explores some of these ideas and demonstrates how from education to economics, there is room for all disciplines in America's largest classroom.

National parks are a quintessential family vacation destination, but what type of learning happens during visits to parks and how might learning be enhanced for families during these visits? In "Lessons Learned from Museums: Family Learning in National Parks," Bourque and Houseal use analogous literature from museums and other free-choice learning spaces to explore research in family learning. Using the personal, sociocultural, and physical contexts from the Contextual Model of Learning (Falk & Dierking, 2000), the chapter synthesizes this literature and proposes recommendations for enhancing family learning in National Park Service (NPS) units and other free-choice settings.

Not only a quintessential family destination, national parks are often associated with environmental education. Powell, Stern, and Frensley wondered if there were consistent learning outcomes to which all environmental education programs for youth should aspire. These authors started with a review of the related literature and engaged dozens of experts across the country to answer this question. "Identifying Outcomes for Environmental Education at National Parks" presents the results of this work, identifying and explaining nine crosscutting outcomes that signify high-quality environmental education at national park sites.

Park programs and publicly available educational materials developed by NPS staff are accessed by tens of millions of people every year. In "Valuing Education and Learning in the National Parks," Marlowe, Bilmes, and Loomis outline their methodology for estimating the direct economic value of NPS educational programs and services. Their approach yields an estimated annual value for NPS educational services of between $949 million and $1.22 billion. This contrasts with an annual budget of approximately $45 million, spent across the NPS on educational activities. Thus, the economic value of NPS educational programs provides a minimum 21–27x return on investment. To illustrate the calculation, the authors share a case study from Golden Gate National Recreation Area (GGNRA).

Despite these detailed and robust valuation efforts, Storksdieck and Falk argue in "Commentary: National Parks as Places for Free-Choice Learning" that the total economic value of a national park is almost impossible to determine. The authors explain that the value of park-based learning cannot be computed because many of the intangible values are difficult to monetize. Park visitors learn about science, society, history, health, and themselves through the activities and people with whom they share their park visit. The authors make the case that we should understand park-based learning not merely as isolated events of supplemental instruction and enrichment, but as integral elements of a person's learning trajectory.

Klondike Gold Rush—family visiting. Courtesy of the National Park Service.

18

Lessons Learned from Museums

Family Learning in National Parks

COLLEEN BOURQUE AND ANA K. HOUSEAL

US national parks are a quintessential family vacation destination, but what type of learning happens during visits to parks and how might learning be enhanced for families during these visits? Lifelong learning, connection to place, and stronger family bonds can all occur as a result of a family's visit to a national park. While individual and collective learning can be difficult to capture, many researchers have devised ways to study learning in various environments.

Over 300 million worldwide visitors are welcomed into 419 park units each year. The percentage of visitors traveling as families is not currently measured, but researchers have estimated that approximately half travel in family groups (Forist, 2003, p. 17). Recognizing this, the National Park Service (NPS) has been exploring ways to enhance lifelong learning for families. This is ultimately critical to the future of parks, since children who visit free-choice settings often return as adults (Beaumont & Sterry, 2005; Moussouri, 2003).

How can the NPS improve family engagement before, during, and after park visits? We argue by considering family visitors' needs and expectations, seeking to better understand their park experiences, and

A version of this chapter originally appeared in Volume 19, Number 1 (2014), of the *Journal of Interpretation Research*.

applying relevant research-based family learning practices. Here we examine factors that influence family learning in free-choice learning environments and present ways to increase family engagement, participation, and learning. We focus our attention on museums as free-choice learning environments due to their learning strategy similarities to national parks. This chapter is centered around this question: What does research literature reveal about the nature of family learning in free-choice environments, factors that influence family learning, and recommendations for improving family learning in free-choice settings?

Parks as Learning Institutions

Our national parks have the potential to play a critical and unique role in fulfilling some of our country's educational needs. Learners typically spend only a small percentage of time in formal learning settings such as schools (Banks et al., 2007). Outside of formal schooling, learning tends to be self-motivated and driven by individual interests, activities, social groups, and surrounding environments. Falk and Dierking (2010) noted that the United States' "vibrant free-choice learning landscape" is a unique and valuable asset to the country's education system (p. 486). Within this learning landscape, national parks and other venues are "contextually relevant and rich places; they are full of real things, situated within relevant contexts" (Falk, 2009, p. 150).

Evidence suggests that free-choice learning experiences, like those provided by the NPS, have the potential to contribute significantly to public literacy. Thus, the NPS has the potential to positively and significantly influence more visitors' learning by focusing on educational improvements that benefit families and lead to further engagement in park efforts and stewardship. In recent years, the *Every Kid in a Park* (EKiP) initiative has provided all US fourth graders and their families free entry into national parks. EKiP is an example of a promising new program that could benefit from an increased focus on family learning to maximize its engagement and impact.

Scope of the Chapter

While the NPS conducts educational programs that are enjoyed by people of all ages, the educational impacts are largely unknown (Brody & Tomkiewicz, 2002). For this review, a few peer-reviewed articles on national park learning were found (e.g., Benton, 2008; Novey & Hall,

2007), but none directly addressed family learning. A parallel body of literature that examines family learning in free-choice settings was used to fill the research gap, as "much of the existing informal education literature is considered to be applicable to park environments" (Brody & Tomkiewicz, 2002, p. 1122). Gross and Zimmerman (2002) directly linked parks and museum settings as comparable venues due to their common audiences, methods of communication, and roles in protecting culturally valuable items.

In 1994, Dierking and Falk recognized that much of the free-choice learning research involved middle-class, Caucasian families and that to ensure generalizability more research on underrepresented populations' use of free-choice settings was needed. A search for literature on under-represented families' use of free-choice settings yielded only a few articles (e.g., Archer, Dawson, Seakins, & Wong, 2016; Gaskins, 2008; Honey, Augare, & Sachatello-Sawyer, 2010). More than two decades after Dierking and Falk noted the lack of research, this is still a scant body of literature.

Four previous literature reviews provided background, context, and common themes in family learning (Adams, Luke, & Ancelet, 2010; Borun, Cleghorn, & Garfield, 1995; Dierking & Falk, 1994; Ellenbogen, Luke, & Dierking, 2004). This chapter builds on those reviews by synthesizing research on the topic published after 2007, research not included in the previous reviews, and relevant information from edited books, non-peer-reviewed articles, and reports. While this chapter focuses on the learning aspects of visits to free-choice settings, it is important to acknowledge that visitor experiences go beyond learning and encompass multifaceted experiences that parallel ones you might find on a national park visit.

RESEARCH ON FAMILY LEARNING

Understanding of family learning has expanded in the last thirty years through a number of research studies. In the mid-1990s, Borun and colleagues embarked upon an extensive research project to improve exhibit design and enhance family learning in Philadelphia-area museums (Borun et al., 1998). The researchers sought to find out whether or not families that appeared to be learning were actually doing so. Cognitive tests were deemed insufficient, since they do not take into account the nuances of family learning. Instead learning was measured qualitatively through interviews. Families were found to have learned from exhibits and there was a relationship between depth of learning and observable behaviors families were engaged in. Researchers also concluded that individual learning was enhanced by input from other family members (Borun, Chambers, & Cleghorn, 1996).

Using prior research and their interview results, the Philadelphia-area researchers created a list of seven characteristics of family-friendly exhibits to incorporate into their museums and test further. According to Borun and Dritsas (1997, p. 180) family-friendly exhibits are

- **multisided** - family can cluster around the exhibit,
- **multi-user** - interaction allows for several sets of hands,
- **accessible** - comfortably used by children and adults,
- **multi-outcome** - complexity of observation and interaction fosters group discussion,
- **multimodal** - appeal to different learning styles and levels of knowledge,
- **readable** - arrange text in easily understood segments, and
- **relevant** - provide cognitive links to visitors' existing knowledge and experience.

Follow-up research using these characteristics to enhance exhibits demonstrated that they did increase active family learning (Borun, Chambers, Dritsas, & Johnson, 1997). These seven characteristics are still being implemented in places of learning today (Borun, 2008).

A decade later, Ellenbogen and colleagues (2004) examined literature written between the mid-1990s and 2004 on family learning in and from museums and similar settings. They noticed three trends. First, converging theoretical perspectives led to shared understandings of what constituted family learning. Second, they noted that researchers were increasingly regarding families as learning institutions in their own right (Ellenbogen et al., 2004). Finally, more rigorous and standardized methodologies were emerging. These three elements revealed that an increasingly cohesive body of research was developing.

A Shift in Theoretical Perspectives

In the early years of family learning research (1980s), most studies were based on behaviorist models of learning that assumed visitors would learn the right material if they were provided with a well-designed exhibit (Falk, 2007). This perspective was institution centered and did not take into account visitors' backgrounds. These behaviorist ideas persist today, though most researchers now agree that a complex suite of factors contribute to free-choice learning (Falk, 2007). Current research is primarily influenced by sociocultural and constructivist theories which advocate a holistic view of learning (Ellenbogen et al., 2004; Falk, 2007; Phipps, 2010). These theories require researchers to examine "the ways in which the family group is situated within the larger social and cultural context" (Ellenbogen et al., 2004, p. S50).

Families as Learning Institutions

Ellenbogen et al. (2004) found that families use free-choice settings as tools or resources to build family identity. Dierking (2010) noted that "the very first learning group a person belongs to is her family and this group is so important that anthropologists, sociologists and social psychologists refer to the family as an educational institution, similar to a museum or school but without the bricks and mortar" (para. 1). Within these unique institutions, family members visiting free-choice settings bring with them identity-related motivations, expectations, and visit plans. These plans are negotiated among members before and during the visit on both a personal level and collectively as a family unit (Moussouri, 2003). In addition, families use shared, memorable experiences to create a common history and narrative. Family learning builds on this history. Experiences in free-choice settings can contribute meaningful, firsthand experiences that serve as building blocks for further learning and identity building.

The Contextual Model of Learning

In addition to family learning, researchers recognized the importance of better understanding factors that influence free-choice learning in general. Falk and Dierking developed an initial framework of contexts that visitors experience before, during, and after their visits to free-choice settings. Their Contextual Model of Learning illustrates how personal, sociocultural, and physical contexts overlap each other and influence visitor learning (Falk & Dierking, 2000). Table 18.1 summarizes this model, which has been expanded to include twelve factors that influence learning. Falk and Storksdieck (2005) explained that "the relative importance of any one of these factors may vary between particular visitors and venues" (p. 747). For example, a family's visit to Yellowstone National Park may be motivated by a parent's memories of visiting the

TABLE 18.1 SUMMARY OF THE CONTEXTUAL MODEL OF LEARNING

Personal context	Sociocultural context	Physical context
Motivation and expectations	Within group social mediation	Orientation to physical space
Prior knowledge and experience	Facilitated mediation by others	Architecture and large-scale environment
Prior interests and beliefs	Cultural background and upbringing	Design of exhibits and content of labels
Choice and control		Advance organizers
		Subsequent reinforcing events and experiences outside the museum

NOTE: Adapted from Falk and Storksdieck, 2005, p. 747.

park as a child, creating a strong personal context component to the visit. Alternatively, a visit to Colonial National Historic Park in Virginia, where a family interacts with costumed park employees demonstrating colonial life, may have strong elements of the sociocultural context of learning.

The following sections of this chapter use the three key components of this model-personal, sociocultural, and physical contexts-as a lens through which to examine recent literature on family learning in free-choice settings. It is important to keep in mind, however, that even though the contexts and factors may be examined separately, they are inseparable. The interaction of the factors is unique for each family and, to a certain extent, for each family member.

The Personal Context

The importance of visitors' prior knowledge and experiences has been emphasized in the free-choice learning literature (e.g., Briseño-Garzón, Anderson, & Anderson, 2007b; Falk & Dierking, 2000; Moussouri, 2003). Family members often share similar beliefs and prior leisure experiences. Briseño-Garzón et al.'s (2007b) study revealed that "participants' interests and what they looked forward to obtaining from the aquarium experience were shaped by particular and personally relevant prior events and knowledge" (p. 87). These events can form a common foundation upon which family learning develops.

In a nature center setting, Zimmerman and McClain (2014) found that "[families] used their prior experiences as learning resources during conversations from three primary sources: (1) experiences outdoors; (2) media such as books, the Internet, and games; and (3) prior experiences at informal education venues-all of which are from informal learning settings" (p. 185). This research emphasized that free-choice and everyday experiences were referenced much more frequently than school experiences in outdoor settings. Therefore national parks serve an important role as a foundation upon which visitors can build knowledge and understanding of a multitude of social, scientific, cultural, and historic phenomena.

The personal context is also shaped by the motivations and expectations of visitors to free-choice learning venues. In McClain and Zimmerman's (2014) study of families at a nature center, "parents, more often than children, referred to memories of everyday experiences during the nature walk conversations" (p. 1018). Families also referenced experiences they had earlier in their nature walk to connect and deepen learning throughout the day. "Reminding was the most common family facilitation process related to prior experiences" in the study (p. 1022). For example, grandparents sometimes communicate memories and wisdom from their past, tapping into and connecting every family members' prior knowledge and experience, which is an important pathway to learning. These tendencies to access prior knowledge helped families learn.

Ethnicity also influences personal context, and we know that "Latinos, Asians, and African Americans are underrepresented among visitors [in national parks] compared to their percent of the U.S. population" (Sheffield & Roberts, 2011, p. 4). Few studies have focused on underrepresented families who engage in free-choice learning opportunities, but the research in this area is growing. Stein, Garibay, and Wilson (2008) noted that because many free-choice learning institutions aim to serve diverse audiences, it is important they consider these visitors' needs "through the lens of the visitors themselves" (p. 180).

There are complex reasons certain groups are underrepresented in parks, but being uncomfortable and unaware of the unspoken rules of free-choice settings may prevent people from visiting. Melber (2006) interviewed Latina mothers at two California natural history exhibit halls. The mothers, who had never visited a local museum, discussed "a fear of not being welcome, not knowing the answers to questions their children may ask, and not feeling that they were knowledgeable enough to appreciate the museum as a learning environment" (p. 37). It is understandable that families without prior visits to national parks may not be comfortable initiating these visits. The personal connections that families make can help bridge gaps in understanding and experience by making learning personally relevant, but it is important to recognize that many complex factors influence the ways families learn and experience their visit.

The Sociocultural Context

Falk and Dierking's (2000) Contextual Model of Learning calls for researchers and practitioners to recognize the social and cultural contexts that visitors bring with them to free-choice settings. The social context is extremely important to families because family members typically interact with each other frequently throughout their visit, often taking turns teaching and learning. Staff and volunteers at free-choice venues may also mediate the experience for family visitors by answering questions and leading programs. In addition, visitors experience free-choice learning through lenses informed by their cultural backgrounds. Free-choice venues present content and experiences through cultural perspectives that are sometimes intentional, but often not. These factors and their influence on family learning are considered in the following sections.

McManus (1992) compared museum-visiting families to hunter-gatherer groups searching for knowledge. Ash (2003) and Falk and Dierking (2000) observed families that split into dyads and triads during their visit and regrouped to share what they learned. Other families stayed together throughout their whole visit. These studies highlight the differences in family social approaches to their visit.

Families often tell researchers they value the collaborative aspects of learning in museums (Allen & Gutwill, 2009; Briseño-Garzón, Anderson, & Anderson, 2007a). Archer and colleagues (2016) found that disadvantaged families visiting free-choice settings valued time together to bond on a special day out, a rarity due to busy work schedules and household management needs. Astor-Jack, Whaley, Dierking, Perry, and Garibay (2007) argued that in order to understand the nature of learning in museums, one must understand the social processes of learning. As a result, researchers have put substantial focus on studying variable family conversation, learning processes, and mediation.

The Power of Conversation

For families, time spent at free-choice settings is typically dominated by conversation, which includes asking questions (usually about specific objects) and sharing knowledge (Falk & Dierking, 2000). Researchers have studied family conversations as a way to gain insight into how and what families are learning, including how frequently families talk throughout their visit.

Borun, Chambers, and Cleghorn (1996) found that the most consistent indicators of learning were in conversations that included analysis, synthesis, and explanation. For example, family conversations often reference previous experiences and are used by families to connect what they are learning to their shared past (Ellenbogen et al., 2004; Falk & Dierking, 2000; Zimmerman & McClain, 2014). The new learning experience then becomes a shared family memory that can be referred to in the future. One reason it might be particularly easy for people to access memories of their visit is that they are usually rare experiences. Crowley and Jacobs (2002) proposed that learning conversations in museums can be powerful foundations for future learning because of this novelty.

Museum Frameworks: Mediating Roles in Family Interactions

Zimmerman and McClain (2014) observed families using different strategies together at a nature center. They found two important intergenerational learning processes happening: (a) prior knowledge of shared experiences "primarily from family experiences in the outdoors, media, and experiences at informal education venues," and (b) family-created participation frameworks that solved family disagreements or included and valued all family members' input (p. 184). "When families created a space to level the hierarchy between elder and child, they negotiated ideas and participated in collaborative idea formation," noted Zimmerman and McClain (2014, p. 184). Families navigate their unique social cultures and also make sense of their surroundings in dynamic ways during their visits.

Szechter and Carey's (2009) study of twenty parent-child dyads in an informal science education center showed that children were the ones who chose exhibits for their families and used hands-on elements more than parents did. Wood and Wolf (2010) also found that parents at children's museums preferred to let children, older than toddlers, initiate the activity across a variety of exhibition types. A number of studies have cited examples of children excitedly sharing with their families what they had learned during a previous school visit (Beaumont & Sterry, 2005; Lyons, Becker, & Roberts, 2010; Moussouri, 2003).

Other studies have focused on parent-child interactions (Astor-Jack et al., 2007; Tuttle et al., 2017). In these, parents have been observed exhibiting what Moussouri (2003) called *spontaneous "teaching" behavior.* Specifically, parents of young children assisted them by posing questions and providing clues and explanations (Moussouri, 2003). Parents and caregivers also used a variety of interaction styles. For example, Szechter and Carey (2009) found that parents described evidence, gave directions, provided explanations, made connections, and elicited predictions. Zimmerman and McClain (2014) found in environmental education settings social learning was present in "parents managing disagreements, families negotiating ideas, and in collaborative idea formation from parents and children together" (p. 189).

Researchers have studied ways that preexhibit instructions might help adult caregivers mediate their children's experiences. Benjamin, Haden, and Wilkerson (2010) found that even brief instructions can improve parent mediation skills. In their study, some caregivers were given suggestions of possible conversational styles and questions. When the prompts were used, there was an increase in children and caregiver's interactions as compared to caregivers who did not receive those instructions. Additionally, Allen & Gutwill (2010) found that "offering parents a structured, co-investigative role in exploring phenomena may significantly enhance families' inquiry" (p. 738). This role helped parents avoid didactic teaching methods or a tendency to delegate simple tasks to children while taking on more difficult tasks themselves. This behavior usually failed to significantly challenge children in a way that enhanced their learning.

Having a role to step into can help parents be better mediators for their children. Tuttle et al. (2017) found that science activities for families that gave adults a navigator role (but not an expert role) and included scaffolding to direct that navigation led to more family conversations and therefore more learning opportunities. Tuttle et al. (2017) recommended the following for educators facilitating family learning experiences:

- Use inquiry-based open-ended activities, which lead to increased intergenerational discussion and interactions, as opposed to closed-ended tasks.

- Notice how facilitators can unintentionally inhibit family conversation and work on ways to help them correct this.

- Use "guiding questions that spell out a clear role for the adult . . . rather than a prescriptive or step-by-step approach to a problem or exhibit" (p. 182).

Besides communicating roles for parents and staff, place-based educational principles can be applied to designing materials that aid families in practicing scientific sense-making skills both in parks and out. Luce, Goldman, and Vea (2017) created resources that could prompt families to practice scientific sense-making skills anytime and anywhere. They use local points of interest in the environment and cue families as to how they might "engage playfully through questioning and exploring" (p. 252). Luce et al. (2017) make the following "design recommendations for facilitating collaborative sensemaking" in national parks and other settings:

- Use local contexts rather than abstract and generalized ones;
- "Prompt exploration of phenomena for which nobody 'knows the answer'" (p. 272); and
- "Prompt exploration of phenomena for which people will likely have multiple, competing ideas" (p. 272).

These are intended to support in-the-moment learning, setting the stage for families to make personal and relevant meaning of their museum visits. This template for mediation often serves parents, but parents do not always lead or mediate experiences in free-choice settings.

Mediator Roles

According to Briseño-Garzón et al. (2007b), roles are often shared among family group members, creating a rotation of roles within a visit. Examples of these include a child taking on the role of exhibit selector, a parent taking on the storyteller role, or a grandparent acting as a questioner. People outside of the family, such as free-choice learning staff and volunteers, often also serve as mediators for families during visits. However, these roles can shift from volunteer to volunteer, from family to family, and from culture to culture.

Falk and Dierking (2000) suggested that skilled staff and volunteers can positively influence and facilitate visitors' experiences. Rosenthal and Blankman-Hetrick (2002) found that staff interpreters who engaged in the right balance of dialogue involving all family members inspired family conversations. However, if interpreters provided too much monologue or too little conversation, there was little indication that visitors were learning. Astor-Jack et al. (2007) made an anecdotal claim that "most interactions between museum staff and the public remain didactic" and advised museum staff to create more participatory family experiences (p. 226).

Apart from the quality of facilitators and volunteers, families' own contexts, norms, and cultures impact who mediates what. Melber (2006) and Stein et al. (2008) cautioned that some family cultures may not encourage children to take on teaching or leading roles. Shouse, Lewenstein, Feder, and Bell (2010) and Stein et al. (2008) also pointed out that nonformal setting (such as park) agendas that encourage children to lead, teach, or challenge their elders' ideas may conflict with families that value didactic approaches or ones in which adults are seen as knowledge holders.

Gaskins (2008) reminded practitioners to avoid the assumption that all cultures share similar theoretical perspectives of how children learn best. For example, while nonformal US educational settings, such as national parks, are often dominated by the theoretical perspective that play leads to learning and that it is appropriate for adults to play alongside their children, play carries different meanings in other cultures (Gaskins, 2008). "Although 'nontraditional' families may enjoy their visits, this enjoyment does not necessarily mean that barriers to inclusion, participation and engagement have been overcome," noted Archer et al. (2016). This is relevant for the NPS to consider because of their desire to welcome and include diverse groups in park experiences (NPS, 2012). Free-choice settings, at their best, can "explore and develop resources and approaches to help support and encourage families to make links between their own backgrounds and experiences and the exhibits" (Archer et al., 2016). It is important that free-choice settings provide accommodations for these potential differences.

The Physical Context

The physical context of a free-choice setting has a strong influence on family visitor experiences. This context encompasses elements such as the venue's website, its architectural layout, seating availability, exhibit order, information displays, and ways visitors might engage in post-visit experiences.

Exhibit Design

Since exhibits may serve as a starting point for family conversations, their design is important. Some exhibits facilitate conversation more than others. In the past few decades, thanks to recommendations from Borun et al. (1998) and others (e.g., Falk & Dierking, 2000), free-choice settings have moved from primarily static exhibits to those that incorporate more interactive, hands-on features. For example, Knutson, Lyon, Crowley, and Giarratani (2016) studied additions to natural history dioramas that were designed to support family learning. They found that "every intervention was successful in increasing dwell time at least three times over baseline" (p. 346). This demonstrates that even older, more expensive exhibits, like dioramas, may be effectively enhanced by adding flexible and changeable prompts and activities.

Astor-Jack et al. (2007) noted progress "particularly in exhibition and program development where there has been some effort to embed socially mediated notions of learning into the design process" (p. 225), which better meets families' needs. Nonscience museums have been creating opportunities for visitors to ask questions and investigate the museum's collection in an inquiry-based way (Allen & Gutwill, 2009). However, it must be noted that there is still room for growth in exhibit design to better meet families' needs (Borun, 2008).

Allen and Gutwill's (2009) research exemplified how games can add structure to inquiry-based, hands-on science exhibits. They compared inquiry games with control conditions for two hundred families and found that their "inquiry games increased the quantity and quality of families' scientific inquiry" (p. 722). In particular a "juicy questions" game was deemed successful at increasing families' inquiry behaviors and encouraged total family participation and collaboration.

Supplemental exhibit materials can also be useful family learning tools. Tenenbaum, Prior, Dowling, and Frost (2010) studied fifty-eight families' visits to a United Kingdom cultural history museum. They found that family learning could be assisted with the support of booklets or activities that involved checking out a backpack designed to guide families through exhibits, even those that featured family-friendly design elements. Families in their study "spent more time at the exhibits when assigned to the booklet and backpack conditions compared to the control conditions" (p. 248) and "children engaged in more historical talk when using the booklets" (p. 241). These findings point to ways in which free-choice settings may supplement existing exhibits without being completely redesigned.

Maker spaces, defined as places where people can create, explore, and learn using materials, are an increasingly popular addition to informal learning. These unique learning environments have not yet been extensively researched. When studying maker spaces in museums, Brahms (2014) found a few key components of successful programs. Among them were "the accessibility and positionality of adult assistance and expertise, the child's intentions and orientations for their own making endeavor, and the priorities and relevant design choices of the learning environment" (p. 94). These factors can be used to ensure more effective family learning in maker spaces.

Reinforcing Events and Experiences

Family learning does not end when visitors walk out of free-choice venues. Recent research supports finding ways to inspire family members to "learn, try or extend the experience at home" and therefore reinforce and expand learning (Foutz & Emmons, 2017, p. 180). Allen and Gutwill (2009) followed up with families after they had left the museum who used their inquiry games during their museum visit and found that 15 percent continued to use them. The authors recommended adding web components to allow families to join citizen science communities to continue their postmuseum experiences. Briseño-Garzón et al. (2007b) concluded that "the learning impact of an informal experience not only resides in the experience itself, but also in the days and weeks following the visit" (p. 87).

Overall, looking at family visits to free-choice settings through personal, sociocultural, and physical perspectives can lead to better understanding of family learning. Despite the absence of research literature on family learning in NPS settings, we posit that ideas from the above studies conducted in a variety of free-choice learning environments can be transferred. The next section will highlights ways the NPS can use these methods to enhance family learning.

CONSIDERATIONS AND RECOMMENDATIONS FROM THE LITERATURE

Even though researchers are still developing and refining ways to document learning impacts, we do know that family learning in free-choice settings can be measured using a variety of approaches. Many go beyond cognitive tests to encompass social and affective elements. We must also remember that, ultimately, learning is personal, is contextual, and takes time (Rennie & Johnston, 2004). Indeed, studies have consistently revealed that families learn through their social interactions. Given these understandings, there are a number of implications, considerations, and recommendations the NPS should entertain in order to improve family learning experiences at NPS units.

A Call to Action (NPS, 2012), a series of recommendations for the NPS, aims to implement practices that will address the themes of *Con-*

necting People to Parks and *Advancing the NPS Education Mission*. However, families are not directly addressed in the document. Though the NPS already reaches young people through school partnerships, EKiP, and Junior Ranger programs, these partnerships could be expanded by connecting youth to parks through more effective multigenerational family programs. We outline our identified considerations and recommendations in an effort to illuminate some key opportunities.

Considerations

First, it is worth noting that families are not the only park visitors. As such, exhibits and programs that encourage social interaction and allow for participation of multiple users will work with families and also school and other adult groups (Borun, 2008; Kiihne, 2008). Next, in exhibit design, research, and visitor interactions, it is valuable for researchers and practitioners to recognize their own cultural lenses before considering visitors' experiences. It is critical that staff "are trained to work with families and groups who are outside the 'typical' visitor profile" (Archer et al., 2016, p. 936). In order to serve visitors who fall into these underserved groups, Shouse et al. (2010) argue that NPS exhibit and program designers must "explore diversity as a positive resource" that adds richness to visitor understandings and perspectives (p. 145). In addition, since changes made in order to serve families better may have unintended effects, careful evaluation of changes will be important.

Recommendations

As we present key recommendations from the literature, recall that the sections of Falk and Dierking's (2000) Contextual Model of Learning are not meant to be separated. Rather, we continue separating them for ease of presentation of the recommendations.

The Personal Context: Maximizing Family Engagement and Learning

In order to fulfill families' needs as communities of learners, the NPS and other organizations are encouraged to consider the following recommendations that take into account the motivations, expectations, knowledge, experiences, interests, and beliefs of visiting families:

- **Visitor agendas** - Seek to further understand families' individual and collective agendas, motivations, and expectations.
- **Prior knowledge** - Help families relate everyday experiences and prior knowledge to what they see and do in NPS units and allow them to investigate issues that interest them (Archer et al. 2016; Moussouri, 2003).
- **Adult learners** - Reach adults who visit as part of family groups as "learners in their own right" (Sanford, Knutson, & Crowley, 2007, p. 148), not just facilitators of their children's experiences (Briseño-Garzón et al., 2007a).

The Sociocultural Context: Maximizing Family Engagement and Learning

Families learn in free-choice contexts by conversing with each other and helping one another. NPS staff and volunteers also serve as mediators of programs and park experiences. The following recommendations will enhance those processes:

- **Social connections** - Develop ways to reward and foster connections between family members, since family groups value the social elements of interacting during their visits (e.g., Moussouri, 2003; Sanford et al., 2007).

Help Parents Facilitate Family Learning

- **Optional help** - Provide instruction for parents to assist them in facilitating their child's learning (Eberbach & Crowley, 2017).
- **Support multiple roles** - Provide information that helps adults quickly recognize their role (Downey, Krantz, & Skidmore, 2010; Falk, 2009) and offer ideas for rotating family roles (e.g., teacher, learner, storyteller; Gaskins, 2008; Leinhardt & Knutson, 2006).
- **Inquiry skills** - Enhance inquiry skills through structured coinvestigative roles for parents (Gutwill & Allen, 2010).

Help NPS Staff and Volunteers Facilitate Family Learning

- **Communication training** - Train facilitators to communicate with and engage all members of multiage families (Falk & Dierking, 2000; Rosenthal & Blankman-Hetrick, 2002).

- **Collaborate** - Create opportunities for novice staff to work collaboratively with knowledgeable mentors (experienced staff/volunteers/experts; Falk & Dierking, 2000).

Cultivate Culturally Relevant Partnerships

- **Model after successful programs** - Engage youth and their families in ongoing programs by valuing reciprocity and involving multiple generations (Honey et al., 2010).
- **Go beyond an invitation** - Attract new audiences by collaboratively developing settings in which "a multitude of cultures feel both welcome and valued and see personal relevance" (Melber, 2006, p. 36; Ng, Ware, & Greenberg, 2017; Stein et al., 2008).
- **Translate and interpret** - Create a welcoming atmosphere by providing translated materials, bilingual labels, and interpretation services for nonnative English speakers (Melber, 2006; Shouse et al., 2010; Stein et al., 2008).
- **Be relevant** - "There is no time like the present for museums to demonstrate their relevance by critically engaging in the political, social, and cultural realities of society today" (Ng et al., 2017, p. 142).
- **Recognize power dynamics** - "Diversity and inclusion work can be shallow or tokenizing-with the potential to re-inscribe and perpetuate white supremacy and oppression, even if the intention is to challenge it-or it can be transformative" (Ng et al., 2017, p. 143).

The Physical Context: Maximizing Family Engagement and Learning

A Call to Action (NPS, 2012) action item number nineteen—*Out with the Old*—addresses changes that should be made in NPS physical contexts. What follows are recommendations for preparing families for visits, extending the benefits, and improving the physical structure of NPS venues:

- **Previsit** - Create a "for families" section on NPS websites. Help parents understand how to take advantage of the site's offerings (Falk, 2009). "Provide an introduction and orientation space, specifically designed for 'first time' visitors" (Archer et al., 2016, pp. 935–936).

· **Post-visit** - Develop opportunities that help families extend their visit conversations and skills (Falk & Dierking, 2000). Link them to other projects beyond the venue.

· **Exhibit characteristics** - Create exhibits that have the seven family-friendly characteristics mentioned earlier (Borun & Dritsas, 1997). The best family exhibits are collaborative and "feature repetition" (Borun, 2008, p. 9; Kiihne, 2008).

· **Supplemental materials** - Supplement existing exhibits with mobile technology, booklets, and/or backpacks that scaffold family learning and promote interaction (Tenenbaum et al., 2010).

CONCLUSION

Instead of approaching the above recommendations as add-ons to already existing efforts, whenever possible the NPS must rethink exhibits, programs, and other interpretive media and experiences from a family perspective (Moussouri, 2003). Parks have welcomed families for over one hundred years and the NPS can take steps to prioritize relevant, engaging, and fun educational opportunities for families as a major component of many visitors' experiences. Packer (2006) concluded that "learning for fun is a unique and distinctive offering of educational leisure experiences" (p. 329), adding that "perhaps one of the most important contributions that museums and other educational leisure settings can make to society is in enabling their visitors to rediscover the joy of learning" (p. 341).

In addition to the recommendations above, it will be critical for the NPS to engage in research regarding all types of learning, including family learning. Quantifying the number of families that visit parks is an important area of research. The fact that there was no NPS research literature to draw from for this review indicates this extreme need. The NPS need not start from scratch; research should be built on the quarter century of work done in museum-like settings. Creating both national and site-specific educational objectives will help to ensure that changes and successes can be measured. Clear education and interpretation objectives and specified "take-home" messages for families should be assessed so that progress in family learning in parks can be tracked over time. An initial direction could be for researchers to look for ways in which the specific nature of learning in NPS settings differs from other free-choice settings.

The NPS might consider adopting an institutional learning framework tied to its mission so that common vocabulary and inventory tools can be used to develop effective and focused exhibits and program evaluations. By creating a "shared institutional understanding of what is meant by 'family learning'" and how to plan to enhance it and recognize it, the NPS can join other institutions that are leaders in learning, such as the Children's Museum of Indianapolis (Foutz & Emmons, 2017, p. 179).

While budgets and logistics may be hurdles, now is the time to consider bolstering the NPS's approach to family visitors as the NPS embarks upon its second century. Free-choice learning has far-reaching beneficial effects on families and these benefits should be shared and promoted. The NPS must find ways to promote its activities, conduct further research, and communicate research findings to engage more Americans. Taking these actions will increase public and policy-maker awareness that the NPS provides exceptional places for lifelong learning and fills an important role in society.

To lead, the NPS must continue to build structures through which its staff can collaborate internally and with other educational and community organizations. Involving families in exhibit and program planning, design, and evaluation will be critically important (Bachman & Dierking, 2010; Moussouri, 2003; Sanford et al., 2007). Overall, creating family-centered opportunities and seeking to engage families in deeper levels of learning and collaboration will enable the NPS to reach the *Call to Action* (2012) goals and will make our national parks and historic monuments exemplary places for family learning.

References

Adams, M., Luke, J., & Ancelet, J. (2010). What we do and do not know about family learning in art museum interactive spaces: A literature review. Retrieved from http://familiesinartmuseums.org/images/pdf/CompleteFLINGLitReview.pdf

Allen, S., & Gutwill, J. P. (2009). Creating a program to deepen family inquiry at interactive science exhibits. *Curator: The Museum Journal, 52* (3), 289–306. doi:10.1111/j.2151-6952.2009.tb00352.x

Archer, L., Dawson, E., Seakins, A., & Wong, B. (2016). Disorientating, fun or meaningful? Disadvantaged families' experiences of a science museum visit. *Cultural Studies of Science Education, 11*(4), 917–939.

Ash, D. (2003). Dialogic inquiry in life science conversations of family groups in a museum. *Journal of Research in Science Teaching, 40*(2), 138–162. doi:10.1002/tea.10069

Astor-Jack, T., Whaley, K. K., Dierking, L. D., Perry, D. L., & Garibay, C. (2007). Investigating socially-mediated learning. In J. H. Falk, L. D. Dierking, & S. Foutz (Eds.), *In principle, in practice: Museums as learning institutions* (pp. 217–228). Lanham, MD: AltaMira Press.

Bachman, J., & Dierking, L. D. (2010). Learning from empowered home-educating families. *Museums & Social Issues, 5*(1), 51–66. Retrieved from http://www.lcoastpress.com/journal.php?id=4

Banks, J. A., Au, K. H., Ball, A. F., Bell, P., Gordon, E. W., Gutiérrez, K. D., & Zhou, M. (2007). Learning in and out of school in diverse environments: Life-wide, life-long, and life-deep learning. Retrieved from http://life-slc.org/docs/Banks_etal-LIFE-Diversity-Report.pdf

Beaumont, E., & Sterry, P. (2005). A study of grandparents and grandchildren as visitors to museums and art galleries in the UK. *Museum and Society, 3*(3), 167–180. Retrieved from http://www2.le.ac.uk/departments/museum-studies/museumsociety

Benjamin, N., Haden, C. A., & Wilkerson, E. (2010). Enhancing building, conversation, and learning through caregiver-child interactions in a children's museum. *Developmental Psychology, 46*(2), 502–515. doi:10.1037/a0017822

Benton, G. (2008). Visitor meaning-making at Grand Canyon's Tusayan museum and ruin. *Curator: The Museum Journal, 51*(3) 295–309. doi:10.1111/j.2151-6952.2008.tb00313.x

Borun, M. (2008). Why family learning in museums? *Exhibitionist, 27*(1), 6–9.

Borun, M., Chambers, M., & Cleghorn, A. (1996). Families are learning in science museums. *Curator: The Museum Journal, 39*(2), 123–138. doi:10.1111/j.2151-6952.1996.tb01084.x

Borun, M., Chambers, M. B., Dritsas, J., & Johnson, J. I. (1997). Enhancing family learning through exhibits. *Curator: The Museum Journal, 40,* 279–295. doi:10.1111/j.2151-6952.1997.tb01313.x

Borun, M., Cleghorn, A., & Garfield, C. (1995). Family learning in museums: A bibliographic review. *Curator: The Museum Journal, 38*(4), 262–270. doi:10.1111/j.2151-6952.1995.tb01064.x

Borun, M., & Dritsas, J. (1997). Developing family-friendly exhibits. *Curator: The Museum Journal, 40*(3), 178–196. doi:10.1111/j.2151-6952.1997.tb01302.x

Borun, M., Dritsas, J., Johnson, J. I., Peter, N. E., Wagner, K. F., Fadigan, K., . . . Wenger, A. (1998). *Family learning in museums: The PISEC perspective.* Philadelphia, PA: Franklin Institute.

Brahms, L. (2014). *Making as a learning process: Identifying and supporting family learning in informal settings* (Doctoral dissertation). University of Pittsburgh. Retrieved from http://d-scholarship.pitt.edu/21525/

Briseño-Garzón, A., Anderson, D., & Anderson, A. (2007a). Adult learning experiences from an aquarium visit: The role of social interactions in family groups. *Curator: The Museum Journal, 50*(3), 299–318. doi:10.1111/j.2151-6952.2007.tb00274.x

———. (2007b). Entry and emergent agendas of adults visiting an aquarium in family groups. *Visitor Studies, 10*(1), 73–89. doi:10.1080/10645570701263461

Brody, M., & Tomkiewicz, W. (2002). Park visitors' understanding, values and beliefs related to their experience at Midway Geyser Basin, Yellowstone

National Park, USA. *International Journal of Science Education,* 24(11), 1119–1141. doi:10.1080/095006-90210134820

Crowley, K., & Jacobs, M. (2002). Building islands of expertise in everyday family activity. In G. Leinhardt, K. Crowley, & K. Knutson (Eds.), *Learning conversations in museums* (pp. 333–356). Mahwah, NJ: Lawrence Erlbaum.

Dierking, L.D. (2010). Laughing and learning. Retrieved from http://www.familylearningforum.org/family-learning/familylearning-overview/what-family-learning.htm

Dierking, L.D., & Falk, J.H. (1994). Family behavior and learning in informal science settings: A review of the research. *Science Education,* 78(1), 57–72. doi:10.1002/sce.3730780104

Downey, S., Krantz, A., & Skidmore, E. (2010). The parental role in children's museums: Perceptions, attitudes, and behaviors. *Museums & Social Issues,* 5(1), 15–34.

Eberbach, C., & Crowley, K. (2017). From seeing to observing: How parents and children learn to see science in a botanical garden. Retrieved from http://upclose.pitt.edu/articles/articles.html

Ellenbogen, K.M., Luke, J.J., & Dierking, L.D. (2004). Family learning research in museums: An emerging disciplinary matrix? *Science Education,* 88(S1), S48-S58. doi:10.1002/sce.20015

Falk, J.H. (2007). Toward an improved understanding of learning from museums: Filmmaking as metaphor. In J.H. Falk, L.D. Dierking, & S. Foutz (Eds.), *In principle, in practice: Museums as learning institutions* (pp. 3–16). Lanham, MD: AltaMira Press.

———. (2009). *Identity and the museum visitor experience.* Walnut Creek, CA: Left Coast Press.

Falk, J.H., & Dierking, L.D. (2000). *Learning from museums: Visitor experiences and the making of meaning.* Walnut Creek, CA: AltaMira Press.

———. (2010). The 95 percent solution. *American Scientist,* 98(6), 486–493. Retrieved from http://web.ebscohost.com/ehost/detail?sid=88ba670f-7be1-435b-90dc876780c03984%40sessionmgr11&vid=4&hid=19&bdata=JnNpdGU9ZWhvc3QtbGl2ZQ%3d%3d#db=aph&AN=54616188

Falk, J.H., & Storksdieck, M. (2005). Using the Contextual Model of Learning to understand visitor learning from a science center exhibition. *Science Education,* 89(5), 744–778. doi: 10.1002/sce.20078

Forist, B.E. (2003). Visitor use and evaluation of interpretive media: A report on visitors to the National Park System. Retrieved from http://nature.nps.gov/socialscience/docs/-Visitor_Use_and_Evaluation.pdf

Foutz, S., & Emmons, T. (2017). Application and adaptation of an institutional learning framework. *Journal of Museum Education,* 42(2), 179–189. doi:10.1080/105986-50.2017.1306663

Gaskins, S. (2008). Designing exhibitions to support families' cultural understandings. *Exhibitionist,* 27(1), 11–19. Retrieved from http://name-aam.org/uploads/downloadables/EXH.spg_08/EXH_spg08_Designing_Exhibitions_to_Support_Families_Cultural_Understandings_Gaskins.pdf

Gross, M.P., & Zimmerman, R. (2002). Park and museum interpretation: Helping visitors find meaning. *Curator: The Museum Journal, 45*(4), 265–276. doi:10.1111/j.2151-6952.2002.tb00064.x

Gutwill, J.P., & Allen, S. (2009). Facilitating family group inquiry at science museum exhibits. *Science Education, 94*(4), 710–742. doi:10.1002/sce.20387

Honey, R.E., Augare, H., & Sachatello-Sawyer, B. (2010). Reciprocal relationships: Effective strategies for working with parents at the Blackfeet Native Science Field Center. *Museums & Social Issues, 5*(1), 117–135. Retrieved from http://www.lcoastpress.com/journal.php?id=4

Kiihne, R. (2008). Following families: From tracking to transformations. *Exhibitionist, 27*(1), 54–60. Retrieved from http://nameaam.org/uploads/downloadables/EXH.spg_08/-EXH_spg08_Following%20FamiliesFrom%20Tracking%20to%20Transformations_Kiihne.pdf

Knutson, K., Lyon, M., Crowley, K., & Giarratani, L. (2016). Flexible interventions to increase family engagement at natural history museum dioramas. *Curator: The Museum Journal, 59*(4), 339–352.

Leinhardt, G., & Knutson, K. (2006). Grandparents speak: Museum conversations across the generations. *Curator: The Museum Journal, 49*(2), 235–252. doi:10.1111/j.2151-6952.2006.tb00215.x

Luce, M.R., Goldman, S., & Vea, T. (2017). Designing for family science explorations anytime, anywhere. *Science Education, 101*(2), 251–277.

Lyons, L., Becker, D., & Roberts, J.A. (2010). Analyzing the affordances of mobile technologies for informal science learning. *Museums & Social Issues, 5*(1), 87–102. Retrieved from http://www.lcoastpress.com/journal.php?id=4

McClain, L.R., & Zimmerman, H.T. (2014). Prior experiences shaping family science conversations at a nature center. *Science Education, 98*(6), 1009–1032.

McManus, P. (1992). Topics in museums and science education. *Studies in Science Education, 20*(1), 157–182. doi: 10.1080/03057269208560007

Melber, L.M. (2006). Learning in unexpected places: Empowering Latino parents. *Multicultural Education, 13*(4), 36–40. Retrieved from http://www.eric.ed.gov/PDFS/EJ759638.pdf

Moussouri, T. (2003). Negotiated agendas: Families in science and technology museums. *International Journal of Technology Management, 25*(5), 477–489. doi:10.1504/IJTM.2003.003114

National Park Service. (2012). A call to action: Preparing for a second century of stewardship and engagement. Retrieved from http://www.nps.gov/calltoaction/PDF/Directors_Call_to_Action_Report_2012.pdf

Ng, W., Ware, S.M., & Greenberg, A. (2017). Activating diversity and inclusion: A blueprint for museum educators as allies and change makers. *Journal of Museum Education, 42*(2), 142–154.

Novey, L., & Hall, T. (2007). The effect of audio tours on learning and social interaction: An evaluation at Carlsbad Caverns National Park. *Science Education, 91*(2), 260–277. doi:10.1002/sce.20184

Packer, J. (2006). Learning for fun: The unique contribution of educational leisure experiences. *Curator: The Museum Journal, 49*(3), 329–344. doi:10.1111/j.2151-6952.2006.tb00227.x

Phipps, M. (2010). Research trends and findings from a decade (1997–2007) of research on informal science education and free-choice science learning. *Visitor Studies, 13*(1), 3–22. doi: 10.1080/10645571003618717

Rennie, L., & Johnston, D. (2004). The nature of learning and its implications for research on learning from museums. *Science Education, 88*(Suppl. 1) S4-S16. doi:10.1002/sce.20017

Rosenthal, E., & Blankman-Hetrick, J. (2002). Conversations across time: Family learning in a living history museum. In G. Leinhardt, K. Crowley, & K. Knutson (Eds.), *Learning conversations in museums* (pp. 305–329). Mahwah, NJ: Lawrence Erlbaum.

Sanford, C., Knutson, K., & Crowley, K. (2007). "We always spend time together on Sundays": How grandparents and their grandchildren think about and use informal learning spaces. *Visitor Studies, 10*(2), 136–151. doi:10.1080 /10645570701585129

Sheffield, E., & Roberts, N. S. (2011). Research literature-then and now. Part 2: Race, ethnicity, culture/Recreation activity and style/Barriers and constraints. Retrieved from http://www.nationalparksonline.org/wp-content /uploads/2011/01/NPPC-Review-of-Existing-NP-Market-Research.pdf

Shouse, A., Lewenstein, B.V., Feder, M., & Bell, P. (2010). Crafting museum experiences in light of research on learning: Implications of the National Research Council's report on informal science education. *Curator: The Museum Journal, 53*(2), 137–154. doi:10.1111/j.2151-6952.2010.00015.x

Stein, J.K., Garibay, C., & Wilson, K.E. (2008). Engaging immigrant audiences in museums. *Museums & Social Issues, 3*(2), 179–195. Retrieved from http://www.lcoastpress.com/journal.php?id=4

Szechter, L.E., & Carey, E.J. (2009). Gravitating toward science: Parent-child interactions at a gravitational-wave observatory. *Science Education, 93*(5), 846–858. doi:10.1002/sce.20333

Tenenbaum, H.R., Prior, J., Dowling, C.L., & Frost, R.E. (2010). Supporting parent-child conversations in a history museum. *British Journal of Educational Psychology, 80*(2), 241–254. doi:10.1348/000709909X470799

Tuttle, N., Mentzer, G.A., Strickler, L., Bloomquist, D., Hapgood, S., Molitor, S., . . . Czerniak, C.M. (2017). Exploring how families do science together: Adult-child interactions at community science events. *School Science and Mathematics, 117*(5), 175–182.

Wood, E., & Wolf, B. (2010). When parents stand back is family learning still possible? *Museums & Social Issues, 5*(1), 35–50. Retrieved from http:// www.lcoastpress.com/journal.php?id=4

Zimmerman, H.T., & McClain, L.R. (2014). Intergenerational learning at a nature center: Families using prior experiences and participation frameworks to understand raptors. *Environmental Education Research, 20*(2), 177–201.

Additional Resources

The *USS Constitution Museum* has experienced the benefits of embracing family learning in exhibits and programs. Their resources and tips are available online at https://engagefamilies.org.

Photo courtesy of the National Park Service.

19

Identifying Outcomes for Environmental Education at National Parks

ROBERT B. POWELL, MARC J. STERN,
AND B. TROY FRENSLEY

INTRODUCTION

The passage of the National Park Service Centennial Act (2016) rein-forced that one of the core missions of the National Park Service (NPS) is to provide education to enhance public awareness, understanding, and appreciation of the resources of the park system through learner-cen-tered, place-based materials, programs, and activities. In this chapter, we focus on environmental education (EE) for youth in national parks and nature centers, which includes programs focused on elements of the nat-ural environment and humans' interactions with them. Recent reviews demonstrate that EE programs have the potential to produce a range of positive outcomes for participants, including increased knowledge, more positive attitudes and behavioral intentions toward the environment, enhanced self-confidence and social interactions, and improved aca-demic motivation and performance, among others (Ardoin, Biedenweg, & O'Connor, 2015; Stern, Powell, & Hill, 2014). In considering the full potential of "America's largest classrooms," the possibilities for impact-ing youth appear nearly limitless.

Is there a consistent set of outcomes to which all EE programs for youth should aspire? At first blush, most experienced environmental educators would likely cringe at the question. Some programs, after all, are intended first and foremost to complement students' achievement of formal curriculum standards. Other programs may seek to develop an emotional connection between a national park and its local community or to develop a sense of environmental stewardship in participants. Still others might be designed to enhance students' social interactions or to build their self-confidence. Moreover, researchers and practitioners alike stress the importance of learner autonomy to achieve their own goals (Fenichel & Schweingruber, 2010; National Research Council, 2009; National Science Foundation, 2008). How could one possibly try to usher all such programs toward a single set of intended outcomes for participants? And why would anyone want to?

Rather than considering the development of a single set of intended outcomes as an operation akin to shoehorning all programs into a one-size-fits-all magic slipper, we consider the effort one of pushing each program to reach its full potential. In our experience as researchers and EE program evaluators, we have seen the full spectrum. On one end we have seen drab lectures or worksheet exercises that might convey facts, but also drain any joy and motivation participants may have had about the subject matter. At the other, we have seen engaging programs that have brought participants the same set of facts accompanied with feelings of elation, awe, solemnity, commitment, curiosity, connection, and/or inspiration. However, if we can step away for a moment from the specific factual subject matter of any particular program, we can begin to see some common ground in what EE programs are actually capable of achieving.

This chapter is a description of our efforts to develop a set of outcomes that is relevant to any program that aspires to label itself as *high-quality environmental education*. We have undertaken this effort as part of a larger study, funded by the National Science Foundation Advancing Informal STEM Learning program and the Institute for Museum and Library Services, in which we are examining the relative influence of different pedagogical approaches to EE on outcomes for participants. This overarching research is intended to help the field to understand why some approaches might work better than others in different contexts. Our process to identify these crosscutting outcomes involved a wide range of experts and practitioners, including academics, program providers, evaluators, and leaders of the Association of Nature Center Administrators (ANCA), the North American Association for Environmental Education (NAAEE), the National Park Foundation (NPF), and the NPS.

REVIEWING OUTCOMES FOR EE

Currently, the goals and outcomes of multiple organizations that provide EE programming reflect some consistency, but a concise and agreed upon list does not currently exist. Where agreement does exist, consensus around how to define and measure shared ideas has yet to be established. We review the predominant outcomes associated with EE and the influential organizations and societal forces that are shaping outcomes for EE in national parks.

Environmental Literacy

One of the primary goals of EE, whether explicit or implied, is to develop students' knowledge, skills, attitudes, and behaviors pertaining to the environment. Each of these should equip individuals to recognize, assess, and then address environmental issues facing their local communities and more broadly support a sustainable global future (Hollweg et al., 2011; Intergovernmental Conference on Environmental Education, 1977). This broad collection of knowledge, skills, attitudes, and behaviors is often collectively called *environmental literacy*. The concept of environmental literacy is largely based on the 1977 Tblisi Declaration, a global UNESCO and UNEP (United Nations Environment Programme) effort to define the goals of EE. According to the Tblisi Declaration, the outcomes of EE programs associated with environmental literacy are as follows:

Awareness-to help social groups and individuals acquire an awareness and sensitivity to the total environment and its allied problems

Knowledge-to help social groups and individuals gain a variety of experiences in, and acquire a basic understanding of, the environment and its associated problems

Attitudes-to help social groups and individuals acquire a set of values and feelings of concern for the environment and the motivation for actively participating in environmental improvement and protection

Skills-to help social groups and individuals acquire the skills for identifying and solving environmental problems

Participation-to provide social groups and individuals with an opportunity to be actively involved at all levels in working toward resolution of environmental problems

Twenty-First Century Skills

Several prominent organizations, including the NPS, the Institute of Museum and Library Services (IMLS), and the Smithsonian Institute, recognize the opportunity of informal settings such as national parks to develop "skills that are critical for addressing 21st century challenges," (National Park System Advisory Board Education Committee [NPS-ABEC], 2014, p. 10) such as social justice, climate change, health care, and effective governance (Fenichel & Schweingruber, 2010; Hollweg et al., 2011; Institute for Learning Innovation, 2007; IMLS, 2009; NPS, 2014; National Parks Second Century Commission, 2009; National Research Council, 2009; National Science Foundation, 2008; Smithsonian Institution, 2010). These twenty-first century skills pertain to a hierarchy of associated knowledge, dispositions, skills, and behaviors related to environmental, science, cultural, health, historical, and civic literacy. For example, scientific literacy includes knowledge/understanding of scientific processes; attitudes toward the importance and validity of science; skills to assess the quality of research findings based on source and approach; and the ability to interpret and apply research findings to solve problems, inform policy, and drive economic development (e.g., IMLS, 2009). Inherently, parks' nationally and globally significant cultural, environmental, and historical resources provide an opportunity for educational programs to support this dynamic skills development.

Positive Youth Development

Based on developmental psychology, educational theory, and stages of moral development (see Dewey, 1899; Kohlburg, 1979; Krathwohl, Bloom, & Masia, 1956; Piaget, 1953) education has the potential to transform the future of today's youth in positive ways. Consequently, today many youth EE programs focus on enhancing a range of positive youth development outcomes (e.g., Carr, 2004; Stern, Powell, & Ardoin, 2011) because these skills and attributes are thought necessary for fostering youth who will excel academically and later in life (e.g., Bowers et al., 2010; Lerner, 2008; Lerner et al., 2005; Seligman, Ernst, Gillham, Reivich, & Linkins, 2009). These positive youth development outcomes include developing social and emotional competence, self-efficacy, self-determination, grit, positive identity, prosocial behaviors and norms, and resiliency, among others (e.g., Catalano, Berglund, Ryan, Lonczak, & Hawkins, 2004), and have been associated with many "best practices" in EE such as inquiry, experiential, and place-based approaches (NAAEE, 2012; Stern el al., 2014).

Educational Standards

Since 2001, the US federal government has required publically supported schools to annually assess student achievement in grades three to eight. Consequently, most EE programs for youth in national parks provide experiences and curriculum that align with state and/or national education standards and support classroom learning through hands-on and direct experiences with nationally and globally significant cultural, environmental, and historical resources. These experiences are thought to assist students in improving academic performance.

Most commonly, the STEM (science, technology, engineering, and mathematics) standards most relevant for EE field trips for youth include *ecological processes*, the *interdependence of organisms*, the *interconnectivity of social and ecological systems*, how *humans may impact the environment*, and how *changes in the environment influence ecosystem function* and *human systems* (e.g., Next Generation Science Standards Lead States, 2013). However, other standards are often addressed as well, including math, social studies, or other science standards. The main point here is that EE, regardless of student grade level, can assist students in the achievement of educational standards.

National Park Service Education

Recognizing the important role that experiences in parks can play, the NPS's *Servicewide Interdisciplinary Strategic Plan for Interpretation, Education and Volunteers* (2014) and the *Vision Paper: 21st Century National Park Service Interpretive Skills* (NPS-ABEC, 2014) prioritize enhancing environmental stewardship (literacy), twenty-first-century skills, and positive youth development, and meeting educational standards in all educational programs for youth. These NPS goals appear particularly relevant for identifying crosscutting outcomes for EE.

OUR APPROACH AND FINDINGS

In light of these existing understandings, we chose to build on this list to create a comprehensive set of target outcomes relevant to a wide range of EE programs. To identify the crosscutting outcomes that EE programming *should* influence, we began by reviewing the scientific literature, including two systematic literature reviews (Ardoin et al., 2015;

Stern et al., 2014) and an unpublished Dephi study that also sought to identify crosscutting outcomes for EE (Clark, Heimlich, Ardoin, & Braus, 2015). Using this list as a starting point, we implemented a systematic approach to directly involve EE experts and practitioners in further identifying and refining appropriate crosscutting outcomes for EE programs for youth.

First, we facilitated a workshop with the NPS-ABEC , ANCA, and NAAEE in late November 2016 in Yosemite National Park. In attendance were leading academics, practitioners, evaluators, and leaders of ANCA, NAAEE, and the NPS. Through a collaborative process following procedures outlined by Fenichel and Schweingruber (2010) and Powell, Stern, and Ardoin (2006), incorporating both the existing literature and input from these subject matter experts (SMEs), we reached preliminary consensus on crosscutting outcomes for youth EE programs provided by the NPS and nature centers.

Following this consensus, we further engaged attendees of the initial workshop, as well as an NAAEE Academic Advisory Group of twelve leading academics and leadership from the National Park Foundation Learning Alliance by iteratively presenting, then receiving feedback, and subsequently refining the list and definitions of these crosscutting EE outcomes. Refining of the list and definitions was completed once general consensus was reached.

Next, we introduce the nine aspirational crosscutting outcomes that resulted from our efforts, with broad definitions for each.

Enjoyment/Satisfaction

Enjoyment, or a general feeling of positive emotions toward an experience, is closely associated with satisfaction, which is perhaps the most basic measure of the success of an EE program. Satisfaction may include the meeting or exceeding of expectations, or a generally favorable assessment of the quality of a program or its instructor(s). We caution that satisfaction should not necessarily be equated with "happiness." An effective program might actually provoke less comfortable emotions that cause deeper learning or reflection (Stern et al., 2013).

Interest/Motivation to Learn

Successful EE programs for youth should inspire an interest in learning. This can be in a general sense, through enhanced curiosity, or in a more specific sense, such as increased interest in learning about science, the environment, or civic engagement. In its strongest sense, an interest in learning can be reflected in whether a participant begins to define themselves as a lifelong learner. Self-identification or self-labeling tends to be more reflective of future behaviors than simple expressions of interest alone (Rise, Sheeran, & Hukkelberg, 2010).

Learning

Learning can be thought of and assessed in multiple ways. While students can be quizzed on specific facts to test their declarative knowledge, many believe that procedural knowledge (knowledge about how things work, how to do things, or how to learn new things) and contextual knowledge (understanding why and under what conditions) are more important (Anderson, 1983). While facts will vary from program to program, we argue that one element should be consistent across any high-quality EE program-that participants should develop an awareness of the interconnectedness and interdependence between human and environmental systems. Programs that fail to provide this knowledge might better be described as natural history or some more narrow disciplinary science than as EE.

Additionally, programs may aspire to develop higher levels of understanding, or contextual knowledge, regarding how and why the relationship between human and ecological systems influence a range of things, including human health, particular wildlife populations, or the climate. Regarding procedural knowledge, programs might target various specific forms-for example, use of a traditional scientific method, some other form of inquiry, or civic engagement skills to influence an environmental issue. Our efforts suggest that participants across all programs, however, should feel empowered to investigate and act upon environmental issues they care about, whatever those issues (or specific actions) might be.

Connection

While learning may be the ultimate outcome for many educators, research suggests that personal, emotional, and affective relationships with place and people are more lasting and powerful in shaping an individual's future attitudes and behaviors (e.g., Chawla & Cushing, 2007; Kals, Schumacher, & Montada, 1999). Quality EE seeks to provide both cognitive and affective connections by providing direct experiences.

These experiences in turn foster appreciation, develop personal relationships, and create meaning (Ardoin, 2006). For many youth, these national park experiences may be the first opportunity to develop a long-lasting connection with park resources or a healthy, natural environment in general. At the most positive end of this spectrum, EE programs have the opportunity to inspire awe in students, leaving them forever changed.

Self-Efficacy

Self-efficacy refers to an individual's beliefs about their ability to learn, organize, and perform specific behaviors to accomplish tasks and goals (Bandura, 1997). In high-quality EE, self-efficacy refers to an individual's belief of personal ability to use critical thinking to solve problems, make a difference in their community, address environmental issues, and influence their environment. In general, positive self-efficacy beliefs are associated with higher degrees of motivation, effort, and persistence, as well as academic performance (Bandura, 1997; Schunk & Pajeres, 2005, 2009). High-quality EE for youth uses experiential and place-based techniques to enhance self-efficacy beliefs (Chawla & Cushing, 2007; Meinhold & Malkus, 2005).

Twenty-First-Century Skills

The partnership for twenty-first-century skills is a relatively new educational framework that seeks to develop content knowledge, specific skills, expertise, and literacies designed to support students' mastery of the four Cs: critical thinking and problem solving, communication, collaboration, and creativity and innovation (IMLS, 2009; Partnership for 21st Century Learning, 2015). EE programs that focus on these skills, especially those that do so in multicultural settings, may enhance students' abilities to address environmental, civic, educational, and cultural challenges today and in the future (e.g., Fadel & Trilling, 2012; Fraser, Gupta, Flinner, Rank, & Ardalan, 2013; Kay, 2010).

Environmental Attitudes

Environmental attitudes, which include an individual's sensitivity, concern, and dispositions toward the environment, are considered to be a key component of environmental literacy (Hollweg et al., 2011; National Environmental Education Foundation, 2015). EE programs focus on fostering positive attitudes toward the environment because, ultimately, individuals' attitudes toward nature influence the way they choose to behave toward the environment (Ardoin, Heimlich, Braus, & Merrick, 2013; Hollweg et al., 2011; Littledyke, 2008; Stern, 2000).

Action Orientation

High-quality EE develops knowledge, attitudes, and skills that enable an individual to make informed decisions regarding future behavioral choices. In particular, EE focuses on developing participants' *intentions* to solve environmental and social problems in their communities or beyond. It is worth noting that while intentions are easier to measure than actual behaviors, they are not always directly correlated with action (Armitage & Conner, 2001).

Action

Because intentions do not always predict future behavior, EE also seeks to provide participants with opportunities to actually employ relevant behaviors. Targeted behaviors may include (a) addressing environmental issues, (b) civic/community involvement, (c) volunteering, (d) recreational choices, and (e) educational/life choices. Broadly, we might categorize these actions as various types of stewardship. Positive environmental stewardship behaviors include recycling, picking up trash left by others, participating in environmental restoration work, volunteering in a range of activities to improve communities, choosing positive recreational activities, and making positive life choices back home. Given the broad diversity of these actions, identifying a single crosscutting behavior is unrealistic for all EE programs. However, our efforts do suggest that participants should feel inspired to *perform* new positive actions geared toward improving their health, their achievement, their communities, or their environments.

CONCLUSION

In this chapter, we raised a provocative question: is there a consistent set of outcomes to which all EE programs for youth should aspire? After reviewing the scientific literature and involving leaders in the field in exploring this question, we think the answer is yes. The crosscutting outcomes identified in this chapter reflect both the roots of our field and future directions. (For more information, see Powell, Stern, Frensley, & Moore, 2019.) If examined holistically, these outcomes are truly achievable by *all* EE programs, irrespective of context and location. Will this require reexamining how EE is done in particular locations? Certainly. Program providers may currently be focused on influencing narrower pieces of knowledge or specific attitudes or behaviors, yet in light of these broader outcomes, program providers should consider how their programs may achieve more. If the EE conducted in "America's largest classrooms" is to reach its full potential, positively and dramatically influencing youth, then critically examining the field's assumptions about how to deliver programming is essential and healthy. With so many pressing issues facing humanity, it is essential that EE continually reassess its goals to meet the educational, societal, and service needs of the twenty-first century.

Acknowledgments

Funding for the study was provided by the National Science Foundation's Advancing Informal STEM Learning program Award 1612416 and in part by the Institute of Museum and Library Services, grant number MG-10-16-0057-16. Any opinions, findings, conclusions, or recommendations expressed in this material are those of the authors and do not necessarily reflect the views of the National Science Foundation.

References

Anderson, J.R. (1983). *The architecture of cognition.* Cambridge, MA: Harvard University Press.

Ardoin, N.M. (2006). Toward an interdisciplinary understanding of place: Lessons for environmental education. *Canadian Journal of Environmental Education (CJEE), 11*(1), 112–126.

Ardoin, N.M., Biedenweg, K., & O'Connor, K. (2015). Evaluation in residential environmental education: An applied literature review of intermediary outcomes. *Applied Environmental Education & Communication, 14*(1), 43–56.

Ardoin, N., Heimlich, J., Braus, J., & Merrick, C. (2013). *Influencing conservation action: What research says about environmental literacy, behavior, and conservation results.* Washington, DC: National Audubon Society.

Armitage, C.J., & Conner, M. (2001). Efficacy of the theory of planned behaviour: A meta-analytic review. *British Journal of Social Psychology, 40*(4), 471–499.

Bandura, A. (1997). *Self-efficacy: The exercise of control.* New York, NY: W.H. Freeman.

Bowers, E.P., Li, Y., Kiely, M.K., Brittian, A., Lerner, J.V., & Lerner, R.M. (2010). The five Cs model of positive youth development: A longitudinal analysis of confirmatory factor structure and measurement invariance. *Journal of Youth and Adolescence, 39*(7), 720–735.

Carr, D. (2004). Moral values and the arts in environmental education: Towards an ethics of aesthetic appreciation. *Journal of Philosophy of Education, 38*(2), 221–239.

Catalano, R. F., Berglund, M. L., Ryan, J. A., Lonczak, H. S., & Hawkins, J. D. (2004). Positive youth development in the United States: Research findings on evaluations of positive youth development programs. *The Annals of the American Academy of Political and Social Science, 591*(1), 98–124.

Chawla, L., & Cushing, D. F. (2007). Education for strategic environmental behavior. *Environmental Education Research, 13*(4), 437–452.

Clark, C., Heimlich, J., Ardoin, N. M., & Braus, J. (2015, October). *Describing the landscape of the field of environmental education.* Paper presented at the North American Association for Environmental Education Annual Conference, San Diego, CA.

Dewey, J. (1899). *The school and society.* Chicago, IL: University of Chicago Press.

Fadel, C., & Trilling, B. (2012). Twenty-first century skills and competencies. In N. M. Seel (Ed.), *Encyclopedia of the sciences of learning* (pp. 3353–3356). New York, NY: Springer.

Fenichel, M., & Schweingruber, H. A. (2010). *Surrounded by science: Learning science in informal environments.* Washington, DC: National Academies Press.

Fraser, J., Gupta, R., Flinner, K., Rank, S., & Ardalan, N. (2013). *Engaging young people in 21st century community challenges: Linking environmental education with science, technology, engineering and mathematics* (New Knowledge Report No. PRCO.106110.06). New York, NY: New Knowledge Organization.

Hollweg, K. S., Taylor, J. R., Bybee, R. W., Marcinkowski, T. J., McBeth, W. C., & Zoido, P. (2011). *Developing a framework for assessing environmental literacy.* Washington, DC: North American Association for Environmental Education.

Institute for Learning Innovation. (2007). Evaluation of learning in informal learning environments. Paper prepared for the Committee on Science Education for Learning Science in Informal Environments. Retrieved from https://www.informalscience.org/sites/default/files/Institute_for_Learning_Innovation_Commissioned_Paper.pdf

Institute of Museum and Library Services. (2009). *Museums, libraries, and 21st century skills.* Washington, DC: Library of Congress. doi:10.1037/e483242006-005

Kals, E., Schumacher, D., & Montada, L. (1999). Emotional affinity toward nature as a motivational basis to protect nature. *Environment and Behavior, 31*(2), 178–202.

Kay, K. (2010). 21st century skills: Why they matter, what they are, and how we get there. In J. Bellanca & R. Brandt (Eds.), *21st century skills: Rethinking how students learn* (pp. xiii-xxxi). Bloomington, IN: Solution Tree Press.

Kohlburg, L. (1979). *The meaning and measurement of moral development.* Worchester, MA: Clark University Press.

Krathwohl, D. R., Bloom, B. S., & Masia, B. B. (1956). *Taxonomy of educational objectives: The classification of educational goals-Handbook II: Affective domaine.* New York, NY: David McKay.

Lerner, R. M. (2008). *The good teen: Rescuing adolescence from the myths of the storm and stress years.* New York, NY: Three Rivers Press.

Lerner, R. M., Lerner, J. V., Almerigi, J. B., Theokas, C., Phelps, E., Gestsdottir, S., . . . Smith, L. M. (2005). Positive youth development, participation in community youth development programs, and community contributions of fifth-grade adolescents: Findings from the first wave of the 4-H study of positive youth development. *Journal of Early Adolescence, 25(1),* 17–71.

Littledyke, M. (2008). Science education for environmental awareness: Approaches to integrating cognitive and affective domains. *Environmental Education Research, 14(1),* 1–17.

Meinhold, J. L., & Malkus, A. J. (2005). Adolescent environmental behaviors: Can knowledge, attitudes, and self-efficacy make a difference? *Environment and Behavior, 37(4),* 511–532.

National Environmental Education Foundation. (2015). *Environmental literacy in the United States: An agenda for leadership in the 21st century.* Washington, DC: Author.

National Park Service. (2014). *Achieving relevance in our second century.* Washington, DC: Author.

National Park Service Centennial Act, HR4680 (2016).

National Park System Advisory Board Education Committee. (2014). *Vision paper: 21st century National Park Service interpretive skills.* Washington, DC: National Park Service.

National Parks Second Century Commission. (2009). *Education and learning committee report.* Washington, DC: National Parks Conservation Association.

National Research Council. (2009). *Learning science in informal environments: People, places, and pursuits.* Washington, DC: National Academies Press.

National Science Foundation. (2008). Framework for evaluating impacts of informal science education projects. Retrieved from http://www.informalscience.org/documents/Eval_Framework.pdf

Next Generation Science Standards Lead States. (2013). *Next generation science standards: For states, by states.* Washington, DC: National Academies Press.

North American Association for Environmental Education. (2012). *Guidelines for excellence.* Washington, DC: Author.

Partnership for 21st Century Learning. (2015). Framework for 21st century learning. Retrieved from http://www.p21.org/about-us/p21-framework

Piaget, J. (1953). *The origin of intelligence in the child.* Routledge & Kegan Paul.

Powell, R. B., Stern, M. J., & Ardoin, N. (2006). A sustainable evaluation framework and its application. *Applied Environmental Education and Communication, 5,* 231–241.

Powell, R. B., Stern, M. J., Frensley, B. T., & Moore, D. (2019) Identifying and developing crosscutting environmental education outcomes for adolescents in the 21st century (EE21). *Environmental Education Research.* doi:10.1080/13504622.2019.1607259

Rise, J., Sheeran, P., & Hukkelberg, S. (2010). The role of self-identity in the theory of planned behavior: A meta-analysis. *Journal of Applied Psychology, 40*(5), 1085–1105.

Schunk, D.H., & Pajeres, F. (2005). Competence perceptions and academic functioning. In A.J. Elliot & C.S. Dweck (Eds.), *Handbook of competence and motivation* (pp. 85–104). New York, NY: Guilford.

———. (2009). Self-efficacy theory. In K.R. Wenzel & A. Wigfield (Eds.), *Handbook of motivation at school* (pp. 35–53). New York, NY: Routledge/ Taylor Francis.

Seligman, M.E., Ernst, R.M., Gillham, J., Reivich, K., & Linkins, M. (2009). Positive education: Positive psychology and classroom interventions. *Oxford Review of Education, 35*(3), 293–311.

Smithsonian Institution. (2010). *Strategic plan: Inspiring generations through knowledge and discovery: Fiscal years 2010–2015.* Washington, DC: Author.

Stern, P.C. (2000). Toward a coherent theory of environmentally significant behavior. *Journal of Social Issues, 56*(3), 407–424.

Stern, M.J., Powell, R.B., & Ardoin, N.M. (2011). Evaluating a constructivist and culturally responsive approach to environmental education for diverse audiences. *Journal of Environmental Education, 42*(2), 109–122.

Stern, M.J., Powell, R.B., & Hill, D. (2014). Environmental education program evaluation in the new millennium: What do we measure and what have we learned? *Environmental Education Research, 20*(5), 581–611.

Stern, M.J., Powell, R.B., McLean, K.D., Martin, E., Thomsen, J.M., & Mutchler, B.A. (2013). The difference between good enough and great: Bringing interpretive best practices to life. *Journal of Interpretation Research, 18*(2), 79–100.

Tblisi Declaration. (1977, October). Intergovernmental Conference on Environmental Education, Tbilisi, Georgia.

Exhibit panel on Alcatraz. Courtesy of Will Elder.

20

Valuing Education and Learning in the National Parks

TIM MARLOWE, LINDA J. BILMES, AND JOHN LOOMIS

The educational role of the National Park Service (NPS) stems from its dual mission to not only preserve the United States' natural and cultural resources unimpaired, but also to provide programming, services, and infrastructure that "extend the benefits of natural and cultural resource conservation and outdoor recreation throughout this country and the world" (NPS, 2016a). Through a combination of park programs and publicly available educational materials developed by NPS, the agency reaches tens of millions of people every year. This includes millions of children and adults who engage in place-based educational programs as well as an even larger public who is able to access NPS materials through websites and original online content. In order to support this understanding, the NPS focuses on place-based education (J. Washburn, personal communication, December 19, 2013), a field that integrates student academic and developmental learning into the context of natural, sociocultural, and historic local environments (Lieberman & Hoody, 1998).

We devised a methodology for estimating the direct value of NPS educational programs and services across the United States. We had to exclude some key difficult-to-measure estimates, such as the NPS's role in its partner organizations' educational efforts and the long-term impact of place-based education on the educational attainment of teenagers. With these exclusions in mind, we believe that our methods yield a likely conservative estimate of the true economic value of the NPS's educational programming.

Our conservative approach yields an estimated annual value for NPS educational services of between $949 million and $1.22 billion. This contrasts with annual expenditures of approximately $45 million (J. Washburn, personal communication, December 19, 2013) across the NPS on educational activities ($38.2 million comes from appropriated funds and $6.4 million in fee funding). The economic value of NPS educational programs thus implies a minimum 21-27x annual return on investment.

To illustrate our approach, we calculated the economic value of the educational programming at one NPS site, Golden Gate National Recreation Area (GGNRA). During the GGNRA 2014–15 fiscal year, the park and its partners produced 1.1 million person-hours of educational programming. Conservatively, we value this on-site educational provision at GGNRA between $12.6 and $15.4 million. Generalized over the entire NPS, we estimate the value of on-site programming to be between $895 million and $1.09 billion. For an in-depth explanation of our methodology, please see Tim Marlowe, Linda J. Bilmes, and John B. Loomis (2020), "Economic Benefits Provided by National Park Service Educational Resources," Linda J. Bilmes and John B. Loomis (2019) and Michelle Haefele, John B. Loomis, and Linda J. Bilmes (2016), *Total Economic Valuation of the National Park Service Lands and Programs: Results of a Nationwide Valuation Study.*

CASE STUDY: THE GOLDEN GATE NATIONAL RECREATION AREA

The GGNRA comprises multiple unique sites that total more than two hundred thousand acres on the west side of the San Francisco Bay, including the Muir Woods, Alcatraz Island, the Marin headlands, the Presidio, Ocean Beach, Fort Funston, and Tomales Bay (NPS, 2016b). With over fifteen million annual recreational visitors, the GGNRA is one of the most visited national park sites, accounting for over 5 percent of visitors nationwide (NPS, 2016c). The GGNRA covers a wide variety of natural landscapes, including redwood forests, beaches, and coastal chaparral, as well as cultural and historical sites. Educational programming at GGNRA is just as diverse as the landscape and includes dozens of different lesson-based youth and adult programs, interpretive tours, and volunteer and internship opportunities. Figure 20.1 illustrates the GGNRA case study and the value of NPS programming at the national level.

Spectrum of On-Site Programming

Less intensive	Interpretive interaction with ranger	Single educational visit to park	Single or multivisit park-based curriculum	Yearlong leadership development programming	More intensive
	Single volunteer experience	Multiday educational visit to park	Multiweek summer camp	Sustained volunteer experience	

Value Estimates of On-Site Programming at GGNRA

Programming Type	Lowest Bound Estimate	Moderate-Realistic Estimate
Youth lesson-based programming	$2.7 million	$4.7 million
Adult lesson-based programming	$900,000	$1.3 million
Interpretive interactions	$9 million	$9.4 million
GGNRA Programming Total	$12.6 million	$15.4 million

National Value Estimates of On-Site Programming Offered by NPS

[1 GGNRA student = 71 national students] x [$12.6–$15.4 million] = **$895 million to $1.09 billion**

Value Estimates of Web-based Programming Offered by NPS

NPS.gov	YouTube	Podcasts	Other Original Content
37.9 million educational sessions, averaging 9.6 minutes = 345 million minutes	118 channels for NPS monuments, historical parks, and battlefields	iTunes and iTunes U courses plus 29 other podcast channels with 442 videos and audio episodes	Webinars and online training content created by NPS cultural resources, partnerships, and science directorate

NPS internet-based interpretive resources were used between 209 and 368 million person-minutes. The value of one-year's use of these education resources is between **$54 and $124 million**.

Scaling Up: Using GGNRA to Make National Estimates

Programming Type	Lower Bound Estimate	Moderate-Realistic Estimate
National total on-site programming	$895 million	$1.09 billion
Internet-based resources	$54 million	$124 million
National Programming Total	**$949 million**	**$1.22 billion**

FIGURE 20.1. Value of NPS Programming.

Figure 20 Data Sources

- Barzanji, T. (2013). Combined data on GGNPC program use [Data set]. San Francisco, CA.

- Barzanji, T. (2014, January). Golden Gate National Recreation Area education inventory. Unpublished.

- Caplan, N. (2015, September 3). Education Lead, GGNRA. (T. Marlowe, interviewer)
- Fonfa, L. (2015, August 18). Education Specialist, GGNRA. (T. Marlowe, interviewer)
- Golden Gate National Recreation Area. (2015, September). GGNRA parks as classrooms student attendance [Data set]. San Francisco, CA.
- Golden Gate National Recreation Area. (2015, September). SIR 12 months GOGA roll-up [Data set]. San Francisco, CA.
- Hitchcock, L. (2015, November 20). Senior Executive Director, YMCA. (T. Marlowe, interviewer)
- Koenen, M. (2015, September 15). Supervisor of Interpretative Facilities and Operations. (T. Marlowe, interviewer)
- Levitt, H. (2013, 31 July). Chief of Communications and Partnerships. (T. Barzanji, interviewer)
- Marine Mammal Center. (2015, November 30). Marine Mammal Center program data [Data set]. Sausalito, CA.
- National Park Service. (2009). National Park Service comprehensive survey of the American public: 2008–2009; Racial and ethnic diversity of National Park System visitors and non-visitors. Retrieved from https://www.nature.nps.gov/socialscience/docs/CompSurvey20082009RaceEthnicity.pdf
- National Park Service. (2015, November 20). Teacher resources. Retrieved from https://www.nps.gov/teachers/teacher-resources.htm
- National Park Service. (2016). National Park Service visitor use statistics. Retrieved November 19, 2016, from https://irma.nps.gov/Stats/
- Pepito, E. (2015, September 19). Associate Director of Youth Leadership, Crissy Field Center. (T. Marlowe, interviewer)
- Pon, D. (2015, June 29). Youth Development Coordinator, Golden Gate National Parks Conservancy. (T. Marlowe, interviewer)
- Roberts, N.S., & McAdams, D. (2015, December 14). (T. Marlowe, interviewer)
- San Francisco Unified School District. (2016). SFUSD 15–16 budget. Retrieved from http://www.sfusd.edu/en/about-sfusd/budget.html

- U.S. News. (2016). High school teacher salary. Retrieved from http://money.usnews.com/careers/best-jobs/high-school-teacher /salary
- Wojcik, L. (2015, December 2). Director, SFUSD Environmental Science Center. (T. Marlowe, interviewer)

References

Haefele, M., Loomis, J., & Bilmes, L. (2016). *Total economic valuation of the National Park Service lands and programs: Results of a nationwide valuation study.* Cambridge, MA: Harvard University Press.

Lieberman, G. A., & Hoody, L. L. (1998). *Closing the achievement gap: Using the environment as an integrating context for learning.* Poway, CA: State Education and Environment Round Table, Science Wizards.

Marlowe, T., Bilmes, L. J., & Loomis, J. B. (2020). Economic benefits provided by National Park Service educational resources. In L. J. Bilmes & J. B. Loomis (Eds.), *Valuing U.S. national parks and programs: America's best investment* (pp. 72–92). New York, NY: Routledge.

National Park Service. (2016a). About us. Retrieved from http://www.nps.gov /aboutus/index.htm

———. (2016b, May 23). Golden Gate National Recreation Area. Retrieved from http://www.nps.gov/goga

———. (2016c). National Park Service visitor use statistics. Retrieved November 19, 2016, from https://irma.nps.gov/Stats/

Photo courtesy of the National Park Service.

21

Commentary: National Parks as Places for Free-Choice Learning

MARTIN STORKSDIECK AND JOHN FALK

INTRODUCTION

In 2016, the 413 US national parks attracted more than 307 million visitors. A small portion of those (about 7 million) were children who attended one or more of the many place-based educational programs on issues as diverse as ecology, geology, conservation, or history and heritage. How and what children learn during these potentially transformative experiences is certainly an important element of the (educational) value-added of national parks. Large-scale studies across the National Park Service (NPS) sites are currently underway to better understand how park-based education programs impact students as learners of science, history, or the environment, while simultaneously shaping their sense of self and relation with history, nature, or society (see Powell, Stern, & Frensley, chapter 19, in this collection). These studies complement a host of smaller scale studies and projects at specific sites (for example, see Bose et al., chapter 10; Davis & Thompson, chapter 5; Houseal, chapter 7; Bourque & Houseal, chapter 18). All of these, collectively, allow us to better document and describe how park visits integrate into a broader ecosystem of educational opportunities. They also provide the foundation for understanding visits to parks not merely as isolated events of supplemental instruction and enrichment, but as integral elements of a child's learning trajectory.

FREE-CHOICE LEARNING

Focusing exclusively on school-based or youth-oriented visits, and not including the 300 million other visitors, underestimates the full contribution that national parks make to the learning ecosystem of the United States and, given the large number of foreign visitors, the world. As described in a report by the National Research Council (NRC; 2009) titled *Learning Science in Informal Environments: People, Places and Pursuits*, and again summarized by Falk and Dierking (2010) in an article titled *The 95% Solution*, most people learn about science or the environment (and most other subjects) after they leave school. They do so as part of a lifelong learning trajectory that is largely determined by the learner and guided by personal, social, and professional interests and motivations. Some of this learning is highly situational and in-the-moment; other learning is structured and closely connected to learner identity and station in life (NRC, 2009, 2015).

From the perspective of a person's lifelong learning trajectory, schooling provides an important initiation and foundation for a person's ability and disposition toward future learning (NRC, 2012). Education and learning during the years preceding early adulthood are, for better or worse, formative in nature, shaping attitudes, knowledge, skills, and identity, and launching young adults. However, all learners, including children and youth, are already learning about the world and being shaped in their motivations, sense of self-worth, and identity by experiences outside of the K–12 classroom (NRC, 2014, 2015).

We do not fully understand the complexity of the multifaceted phenomenon of free-choice learning during a visit to a national park. We do know, however, that analogous to museum visits (Falk & Dierking, 2014, 2018), park patrons learn about the park through the activities that the visitor takes part in and the people with whom they do these activities. This learning process is strongly influenced by the degree to which individuals have choice and control over their visit, by whether they experience joy and satisfaction, and by the degree to which a visit is aligned with their sense of self-related needs (Bond & Falk, 2012; Falk, 2009; Falk & Storksdieck, 2005).

FREE-CHOICE LEARNING'S VALUE IN THE NATIONAL PARKS

Marlowe, Bilmes, and Loomis present their process for economically valuing national parks in chapter 20. In short, their argument places economic value on many of the measurable resources and educational offerings of the NPS. While these estimates and valuations are immensely useful in their applications to park decision-making and funding, we would argue that their very modest estimate misses a fundamentally important piece of the learning experience. In this volume and in previous publications by the authors and their colleagues (e.g., Haefele, Loomis, & Bilmes, 2016), the researchers have been able to place monetary value on a wide array of park-related resources, including the following:

- perceived value of the geographical holdings of the NPS (the land, water, historic sites, etc.);
- use and option value of visiting a park and taking part in its educational or interpretive programming;
- existence and bequest value that US residents assign to the NPS;
- economic benefits of educational programming in the Golden Gate National Recreation Area, extrapolating to NPS's educational programs nationally (amounting to a stunning $30 billion); and
- recreational use of national parks (Neher, Duffield, & Patterson, 2013).

Despite these detailed and robust valuation efforts, the *total* economic value of a national park is almost impossible to determine with certainty, in part because many of its intangible values are not traded in markets and are hence difficult to monetize (Freeman, 2003). While direct economic benefits from business activities, and sometimes ecosystem services, can be modeled and estimated using a variety of methods, determining intangibles poses ethical and methodological difficulties. Not least of these is the question of whether the very act of placing a dollar value on a public good may reduce it in our mind to a consumer good and thereby change dramatically the quality of the thing we intend to measure (Kelman, 1981; Pearce & Turner, 1990; Storksdieck, 1998).

Cognizant of the dangers inherent in this valuation process, a number of scholars (see Champ, Boyle, & Brown, 2003; Clawson & Knetsch, 1966; Hanemann, 1991) have developed valuation methods to address the more challenging-to-quantify measures. In addition to the value derived from direct use of a park, these methods contribute to a more complete valuation; they include the option value of future visits (the amount of money people would be willing to spend to keep the option for themselves of visiting a national park), the bequest value or the benefit to individuals for knowing that future generations may visit the park, and the existence value, often measured as the amount of money people would be willing to pay to sustain a park, even under the condition that they or someone they care about would never be able to visit it. These methods capture a breadth of measures, but their shortcomings are far-reaching as well. For example, Haefele and colleagues (2016) estimate these approaches do not include the value of the NPS to the considerable number of foreign visitors or residents in other countries, who may value Monument Valley or the Grand Canyon just as much as US residents. The authors further caution that their estimate does not capture the role of the NPS in providing important ecosystem services and other aspects of value that surveys cannot capture easily or with confidence.

When placing value on free-choice learning in the NPS and beyond, researchers run up against similar methodological limitations. Surveys, a reasonable means for collecting these data, may not specifically ask respondents about key values implicit in free-choice learning: the incidental learning by simply being there or by studying the park map and other forms of interpretive information; the learning about others with whom one visits; the potential learning that occurs in preparation or as follow-up to the visit; or the sense of identity—regional, national, and international—that a visit can support. By not asking specifically about these aspects of the visit, these surveys may simply not capture their value. In fact, much like museum visits, it is highly likely that many visitors might not be fully aware of the free-choice learning benefits they reap because they do not have an adequately deep and rich understanding of the nature of learning. Because they may not have the language to describe the full extent of their learning, to others or even themselves, many visitors are likely to not fully appreciate the actual learning benefits they accrue from a park visit. In a study by the authors on the learning that visitors derived from a science center visit (Falk & Storksdieck, 2005), it was found that respondents described many ways in which they learned from the California Science Center's *World of Life* exhibit gallery, yet denied having learned since they confounded learning with education and education with schools. That is, visitors to the science center were not cognizant of their own free-choice learning, despite objective measures to the contrary.

CONCLUSION

In museums, national parks, and society as a whole, free-choice learning is often overlooked or undervalued and underappreciated. Yet, it is an important component of lifelong, life-wide, and life-deep learning (NRC, 2009). Free-choice learning experiences form the foundation of the public's learning, and settings like parks represent key elements of the public's overall learning ecosystem (Falk, 2017; Falk et al., 2017; NRC, 2014, 2015). In fact, free-choice learning, in all its aspects, may play an equal if not more impactful role in shaping what we know, what we can do, and who we are than formal education (Falk & Dierking, 2010; Falk & Needham, 2013). Even though we may feel a need to estimate its economic impact in dollars and cents, we would be well advised to consider the many benefits that visitors collectively derive from their visit-related, free-choice learning when describing the full economic and educational value of our national parks.

References

Bond, N., & Falk, J.H. (2012). Who am I? And why am I here (and not there)? The role of identity in shaping tourist visit motivations. *International Journal of Tourism Research, 15*(5), 430–442.

Champ, P.A., Boyle, K.J., & Brown, T.C. (2003). *A primer on nonmarket valuation.* Norwell, MA: Kluwer.

Clawson, M., & Knetsch, J. (1966). *Economics of outdoor recreation.* Baltimore, MD: John Hopkins University Press.

Falk, J.H. (2009). *Identity and the museum visitor experience.* Walnut Creek, CA: Left Coast Press.

———. (2017). *Born to choose: Evolution, self, and well-being.* New York, NY: Routledge.

Falk, J.H., & Dierking, L.D. (2010). The 95% solution: School is not where most Americans learn most of their science. *American Scientist, 98,* 486–493.

———. (2014). *The museum experience revisited.* Walnut Creek, CA: Left Coast Press.

———. (2018). *Learning from museums* (2nd ed.). Lanham, MD: Alta Mira Press.

Falk, J.H., & Needham, M.D. (2013). Factors contributing to adult knowledge of science and technology. *Journal of Research in Science Teaching, 50*(4), 431–452.

Falk, J.H., & Storksdieck, M. (2005). Using the Contextual Model of Learning to understand visitor learning from a science center exhibition. *Science Education, 89,* 744–778.

Falk, J.H., Storksdieck, M., Dierking, L.D., Babendure, J., Canzoneri, N., Pattison, S., . . . Palmquist, S. (2017). The learning SySTEM. In R. Ottinger (Ed.),

STEM ready America (pp. 2–13). Flint, MI: Charles Stewart Mott Foundation. Retrieved from http://stemreadyamerica.org/the-learning-system/

Freeman, A. M. (2003). *The measurement of environmental and resource values: Theory and methods* (2nd ed.). Washington, DC: Resources for the Future.

Haefele, M., Loomis, J., & Bilmes, L. (2016). *Total economic valuation of the National Park Service lands and programs: Results of a survey of the American public* (Faculty research working paper series; RWP16–024). Retrieved from https://research.hks.harvard.edu/publications/getFile.aspx?Id=1395

Hanemann, M. (1991). Willingness to pay and willingness to accept: How much can they differ? *American Economic Review, 81*(3), 635–647.

Kelman, S. (1981). Cost-benefit analysis: An ethical critique. *Regulation, 5*(1), 33–40.

National Research Council. (2009). *Learning science in informal environments: People, places and pursuits.* Washington, DC: National Academies Press.

———. (2012). *Education for life and work: Developing transferable knowledge and skills in the 21st century.* Washington, DC: National Academies Press.

———. (2014). *STEM learning is everywhere: Summary of a convocation on building learning systems.* Washington, DC: National Academies Press.

———. (2015). *Identifying and supporting productive STEM programs in out-of-school settings.* Washington, DC: National Academies Press.

Neher, C., Duffield, J., & Patterson, D. (2013). Valuation of National Park System visitation: The efficient use of count data models, meta-analysis, and secondary visitor survey data. *Environmental Management, I*(3), 683–698.

Pearce, D. W., & Turner, R. K. (1990). *Economics of natural resources and the environment.* Baltimore, MD: Johns Hopkins University Press.

Storksdieck, M. (1998). *Das McCloy Programm: Idee und Ideologie. Drei kritische Anmerkungen eines Teilnehmers.* [The McCloy program: Idea and ideology—Three critical remarks of an alumnus]. In S. Lorenz & M. Machill (Eds.), *Transatlantik: Transfer von Politik, Wirtschaft und Kultur* [Transatlantic: Transfer of politics, economy and culture] (pp. 463–480). Opladen, Germany: VS Verlag for Social Sciences.

Afterword

In 1916, Congress passed and President Woodrow Wilson signed the law establishing the National Park Service (NPS). The NPS Organic Act states, in part:

> To promote and regulate the use of the Federal areas known as national parks, monuments, and reservations hereinafter specified by such means and measures as conform to the fundamental purpose of the said parks, monuments, and reservations, which purpose is to conserve the scenery and the natural and historic objects and the wild life therein and to provide for the enjoyment of the same in such manner and by such means as will leave them unimpaired for the enjoyment of future generations.

For the past one hundred years, leaders of the NPS have interpreted this law of conservation and historic preservation as a mission. While serving as director, I especially focused on the phrase "in such manner and by such means" as a broad mandate to use the magic of the national parks to inspire and educate a new generation of park visitors, advocates, and supporters. I knew that if we do not engage each succeeding generation, how could we expect the parks to remain unimpaired for future generations? Or more broadly, education in the national parks deepens citizen understanding of history, the environment, the ideals of the nation, and our individual responsibility to contribute to society.

The NPS has been collaborating with schools and educators for decades, and during my tenure as director, we expanded the effort significantly with a Strategic Plan, the *Every Kid in a Park* pass, and new programs on climate change, civil rights, and slavery as the cause of the Civil War. We sought out the untold stories of America, women, Latinos, Asian Americans, African Americans, Pacific Islanders, and the LGBTQ community. With classroom teachers, we curated and improved curriculum materials and established a teacher's portal on the NPS website, giving them access to hundreds of resources for students, and through the National Park Foundation, raised millions for school field trips to parks. Through the National Park System Advisory Board Education Committee, led by Dr. Milton Chen, we engaged education professionals from around the nation. And finally, on December 16, 2016, President Obama signed into law the National Park Service Centennial Act, which includes this section:

> *The Secretary shall ensure that management of System units and related areas is enhanced by the availability and use of a broad program of the highest quality interpretation and education. The term 'education' means enhancing public awareness, understanding, and appreciation of the resources of the System through learner-centered, place-based materials, programs, and activities that achieve specific learning objectives as identified in a curriculum."* Public Law No: 114–289 Section 301

In 2017, Clemson University Professor (and my former science advisor) Dr. Gary Machlis and I coauthored a book titled *The Future of Conservation: A Chart for Rough Water*. From our collective experience of over eighty years working with national parks, we asserted that the future of conservation and the national parks, as well as the planet, requires collaboration among multiple sectors of society, from classic nature conservation to education to environmental justice. We argue for a communication strategy that looks to underlying causes and values. Our assertions are confirmed by the authors of this book, as they detail research with learners to understand preexisting knowledge and attitudes around complex issues such as climate change, mental health, environmental conditions, and diversity. Such research supports our assertion that sustained listening and unbiased surveys give us the opportunity for serious reflection. Dr. Machlis and I also suggest this is the time for innovative experimentation, and the case studies (see chapters 3, 6, 8, 11, 12, 15, and 16 in this collection) of programs, such as the Teton Science School, Journey through Hallowed Ground, and high school trades training in New York City, all demonstrate strategic intent and bode well for the future of conservation. Dr. Machlis and I wrote of the value of firsthand contact with nature and history, and the authors of chapters 13, 20, and 21 convert that value into economic return on investment, public health improvements, and the creation of lifelong learners who are interested in and engaged in the world around them. From this rootstock will grow the next generation of conservation leaders, armed with both passion and innovative approaches, unbound by the silos of the past. The resources in this book demonstrate the effectiveness of innovative prototypes that I hope will be shared widely, practiced, and refined.

This book asserts that our nation's citizenry is better informed and better prepared for the challenges that lie ahead when they experience history on sacred ground, nature in wild places, and stories that are authentic and relevant. In 2016, the nation celebrated the centennial of the National Park Service, reaffirming the role our national parks play in the American psyche—as maker of family vacation memories; as keeper of our cultural heritage; as reminder of our failures, challenges, and sacrifices; as scenic wonderland; as home for wildlife and what Wallace Stegner termed "the geography of hope." We now have, as established by this book and the new mandate, one more role for the national parks: *America's Largest Classroom.*

Jonathan B. Jarvis
18th NPS Director, retired
Executive Director
Institute for Parks, People and Biodiversity
University of California, Berkeley

Index

ABEC, 249, 250
Acadia National Park, 37, 44; Schoodic
    Point SeaWatch, 93
Achieving Relevance in Our Second Century
    (National Park Service), 18, 171–73
Adams-Rodgers, Lois, 168
Advancing the National Park Idea
    (National Parks Conservation
    Association), 16
African Americans, xvii, 33, 141, 147, 214,
    227, 271
Alabama, 33, 136
Alaska, 16
Alcatraz National Historic Landmark, xvii,
    258, 260
America's Best Idea (Ken Burns documen-
    tary series), iv
American Indians, xvii, 2, 30, 37–46
American Revolution, 30
American Samoa, xix
Amskapi Pikuni people, *see* Blackfeet
    Indians
Anhinga Trail, 58
Annual Cherry Blossom Festival, 60
anthropocentric viewpoint, 121, 122
anthropogenic impacts, 153, 161
Antietam, Battle of, 30
Appalachian Mountains, 100, 102, 106
Appalachian Trail, 102
Applied Research Northwest, 178
apps, xvi, 133, 154, 156–58, 160

Arches National Park, 37
Arlington House National Historic Site, 29
Armstrong, Neil, 10
Army, U.S., 1, 5
artifacts, xiv, 122, 206, 207
Asian Americans, 210, 271
Asians, 227
Association for Conflict Resolution, 55
Association of Independent California
    Colleges and Universities, 157
Association of Nature Center
    Administrators (ANCA), 246, 250
augmented reality, xvi, 112, 152, 154,
    156, 161
Azevedo, Paula Cristina, 169, 201–214

Backcountry Horsemen of North Carolina,
    102, 103
Badger-Two Medicine area, 39
Baker Massacre, 41
Battery Park (Manhattan), 197
Beaver Meadows Visitor Center, 55
benefits of nature experiences, 127–28, 132,
    237, 239, 259, 267–69
Benfield, Jacob, 112, 127–33
Bethune, Mary McLeod, 202, 206–11
Bethune Council House National Historic
    Site, 204–11
Big Catatoochee Mountain, 92
Bilmes, Linda J., xvii, 218, 259–63, 266–67
Biltmore Mansion, 100

Bio Blitzes, xvi, 16, 93
biodiversity, xiv, xix, 140, 172
biology, 116, 154, 161, 185
biomes, 152
biophilia hypothesis, 115, 121
birds, xvi, 61, 89, 98
Biscayne Bay National Park, 54, 56
Black History Month, 147
Blackfeet Indians, xvii, 2, 37–46
Blankman-Hetrick, J., 231
Blimes, Linda
Bloomberg, Michael, 196
Bose, Mallika, 112, 127–33
Bourque, Colleen, 217, 221–39
Boys & Girls club, 168
Brooks, Geraldine, 30
Brown, John, 30
Brown v. Board of Education of Topeka
    (1954), 33
Bryant, Harold C., 7
Bryce Canyon National Park, 37, 115
Buffalo Soldiers, xvi
Buford, General John, 29, 30
Burnell, Esther and Elizabeth, 7
Burns, Ken, iv
Bursztyn, Natalie, 112, 151–61

Cabrillo National Monument, xvi
Cadillac Mountain, 44
California, xvi, 227. See also specific
    national parks; Bay Area, xiv;
    Community Colleges, Academic Senate
    of, 157; Science Center, 268; State
    University, xvi, 150, 157; University
    of, 157
Call to Action, A (NPS), 17, 171–72 234,
    237, 239
Camp David, 129
campfire talks, 8
Canyonlands National Park, 48
Carbaugh, Donal, xvii, 2, 37–47
Career and Technical Education (CTE) high
    schools, 196
Cascades Climate Challenge, 66
case studies, xix, xx, 74–79, 120, 152, 169,
    202; Freedom Rangers program,
    141–46; Golden Gate National
    Recreation Area, 218, 260; Grand
    Canyon Expedition, 156–60
Castle Clinton National Monument, 197
Catoctin Mountain National Park (CATO),
    112, 126, 127–32
Center for Western Weather and Extremes,
    101

Chaco Culture National Historical Park,
    122
Channel Islands, xvi
Chavez, César, National Monument, iv
Chen, Milton, xiii–xviii, 17, 196, 272
Children's Museum of Indianapolis, 239
Chinampas Basin, 155
Christianity, 39, 40
citizen science, xvi, 49–50, 75, 83–93, 176,
    234
civic engagement, 16, 167, 251
Civil Rights movement, 202, 210, 214, 271
Civil War, 29–34, 62, 271; sites commemo-
    rating, 33, 112, 137, 138, 141, 144, 146
Civil War Trust, 33
clean air, xiv, 10
Clemson University, 272
climate change, 16, 17, 49, 53–67, 153,
    161, 168, 248; engagement in concerns
    about, xx, 50, 61, 65–67; free-choice
    learning about, 61–63; impacts of,
    56–61, 167, 172; place-based education
    about, xvi, 55–56, 63–67
Climate Change Academies, xvi
Coble, Theresa, 1, 23
collaboration, xvi, 15, 18, 59, 167. See also
    citizen science, partnerships; on
    curricula, 12, 184, 197; on research, 54,
    85, 202; with universities, 158, 202
collaborative learning, xvi, 18–19, 112,
    155–56, 158, 169; family, 228–33; on
    field trips, 140, 158–61, 206–8, 212; in
    schools, 158, 169, 198–99, 202–4, 206,
    213, 214
college students, xviii, 116, 152, 154, 158,
    205. See also specific colleges and
    universities; field research by, xvi, 50,
    99–100, 105–7; motivation of, 245–47,
    252; off-campus learning experiences of,
    99, 116
Colonial National Historic Park, 226
Colorado, 11, 54, 55, 58, 83, 90, 116
Colorado River, 123, 152
Colorado Rockies, Front Range of, 115
Columbia (Missouri) School District, 77
communication, 24, 37, 62, 159, 177, 272;
    climate change, 57, 65; in collaborative
    process, 190, 231, 223; development of
    skills in, 236, 252; scientific, 107 ;
    technologies for, 12, 85, 132, 161, 179
community engagement, 17, 76
competency-based learning, 74
Confucius, 208
Congress, U.S., xix, 8, 16, 271

Connecting People to Parks and Advancing the NPS Education Mission (NPS), 235
conservation, 56, 112, 198, 271, 272 ;
benefits of, 17, 127; education on, 9,
265; of resources, 66, 259
conservation psychology. *See* environmental
psychology
Constitution, U.S., First Amendment, 205
Contextual Model of Learning, 217, 225,
228, 235
Cosby Knob backcountry shelter, 102
Costigan, Heather, 112, 127–33
crosscutting, 90, 92, 217–18, 224, 246,
249–50, 253–54; interdisciplinary,
86–87, 87, 89, 91
cultural discourse analysis, 2, 37
cultural diversity, xiv, 187
curriculum, 168, 184–85
Cuyahoga River, 185
Cuyahoga Valley Environmental Education
Center, xv, 168, 183–87
Cuyahoga Valley National Park, 168, *182,*
183–87; Conservancy for, 184–86; Pura
Vida program at, 168, 189–92

Dahlen, David, 13
Darwin, Charles, 85
Davis, Shawn, 49, 53–67
Death Valley National Park, 153
deep listening, xvii, 39, 42
deeper learning, xiv, xvi, 74, 250
Deferred Action Childhood Arrivals
(DACA), 191
Denali National Park, 153
Dephi, 250
design thinking, 74, 80
Dexter Avenue Baptist Church
(Montgomery, Alabama) *136*
Disciplinary Core Ideas (DCIs), 86–87, 89,
91, 92
District of Columbia. *See* Washington, D.C.
Ditch Creek, 75
diversity, 111, 186, 191, 235, 237, 253. *See
also* biodiversity; cultural, 14, 187; in
environmental education, 137, 158, 213
Dodana Manor, 32
Duke University, 50; Great Smoky
Mountains Rain Gauge Network,
97–98, 100, 101–5, 107
Dutton, Clarence, 151

Eagle Cliff, 121
economy, xiii, 73–74, 79, 167, 172, 248;
value of educational programs in

National Parks to, xvii, 17, 140,
217–18, 259, 259–60, 266–69, 272
ecosystems, 75, 79–80, 92, 167, 190, 249;
educational, xiv, 18, 49–50, 265–69;
immersive programs on, xv, xvii;
impacts of climate change on, 54, 57, 80
ecotourism, 50
Education and Environment (Lieberman), iv
Education Initiative Symposium, 14
educational programs, xix, 1, 5, 7–8, 12,
222, 248–49. *See also specific learning
approaches;* case study of, 260;
economic value of, 218, 259–60, 267;
historical, 8–11, 62, 206; partnerships
in, 14, 184, 187
Eisenhower, Dwight D., 30
elementary schools, 80, 202, 205, 211
Emerald Lake, *114*
endangered species, xvii, 10, 152
energy, xvii, 89; clean, xiv, 60
engagement, xiv–xvii, xix–xx, 16–18, 56,
67; in citizen science, 89, 93; civic, 16,
167, 251; in climate change concerns,
xx, 50, 61, 65–67; in education, 14, 32,
34, 49, 54, 64, 73–80, 198–99, 250;
experiential, 116–17 [earlier]; family,
217, 221–22, 224, 227, 231–33,
235–39; on field trips, 153–61; with
historical content and context, 24–26,
137–47, 205, 207, 212; with nature,
psychology of, 120–22; in partnerships,
167–80, 184–87, 189–92, 237; in
science learning, 86–88, 90–92
Environment & Energy News, 175
environment, xvi, xix, 18, 25, 64; challenges
to, xiv. *See also* climate change; literacy
about, 57, 247, 249, 252; nurturing
appreciaition of, 6, 127; stewardship of,
198, 246, 249, 253
environmental education (EE), 9–10, 61,
63–65, 77, 84, 111, 117, 140, 158,
245–54; curriculum for, 168, 184–85;
for families, 217, 221–39; partnerships
and teams in, 186–87, 190, 197–98;
technological approaches to, 154–56,
161
environmental psychology, xvii, 116–20,
122–23
Erie Canal, xv
ethnicity, 15, 17, 112, 137, 158, 174, 202,
227
ethnography, 37, 202
evaluation, 11–13, 15, 16, 179, 210, 235,
239

Everglades National Park, 54, 56, 65
Everhart, Bill, 9
Every Kid in a Park (EKiP) program,
   176–79, 222, 235, 271
exhibits, 7, 8, 24, 138, 141, 258; family-
   friendly design of, 223–24, 226, 227,
   229–33, 235, 238–39
Exit Glacier trail, 53–54, 58
Expedition Yellowstone!, 84, 86, 90
expeditions. *See also* field trips; wildlife, 50,
   72, 78
experiential learning, 7, 63, 73, 116–17,
   153, 252. *See* field trips, place-based
   learning

Facebook, 151
Fahey, John, 199
Falk, John H., 18, 61, 218, 222, 225, 228,
   231, 235, 265–69
family learning, 217, 220, 221–39;
   collaboration in, 237; context of,
   225–28, 232, 235–38; exhibit design
   for, 232–34, 238; mediating roles in,
   229–32; culturally relevant partnerships
   in, 237; reinforcing events and
   experiences in, 234
Feder, M., 232
Fenichel, M., 250
field research, xvi, 8, 50, 97–108, 198
field trips, xiv, 7, 12, 153, 177, 197, 249; to
   historic sites, 203, 211, 212; virtual and
   augmented reality, xiv, 112, 153–56,
   158, 161
Fifth Essence, The (Tilden), 9
Find Your Park/Encuentra Tu Parque
   campaign, 176–77, 179
First Amendment, 205
First Peoples, 2, 38
Flattop Mountain, 121
Florida, 54, 100
forests, xv, xvi, 102, 105, 129, 184, 197,
   260; climate change impacts on, 58–59;
   invasive species in, 54, 56
Fort Funston, 260
Framework for K-12 Science Education, 83,
   86, 88, 93
Franklin, Benjamin, 85
Franklin, John Hope, 14
Fraser, Loran, 14
Frederick Douglass National Historic Site,
   iv
Fredericksburg and Spotsylvania National
   Military Park, 31
Freedman, Teddi, 168

Freedom Rangers, 137–47, *139, 140,*
   145–46
Frensley, B. Troy, 217, 245–54
Future of Conservation: A Chart for Rough
   Waters, The (Machlis and Jarvis), 272

Gabbard, Larry, 198
GCX apps, 156–58, 160
gender, 130, 146, 158, 175, 202, 204
Generation Z, 167
geography, xix, 154, 161, 175, 185
geology, xvi, 157–60, 265
George Lucas Educational Foundation, 17
George Mason University. 202
George Washington University, 17
geoscience, 112, 153, 154, 157–58, 160–61
Gettysburg National Military Park, iv, 2,
   29–30, 33
Glacier Basin campground, 55–56
Glacier National Park, xvii, 36, 37, 39, 41,
   46
Golden Gate National Recreation Area
   (GGNRA), 218, 260, 267
Goldman, S., 231
Goode, Richard, 112, 151–61
Google, xvi
Governors Island National Monument, 196
GPS technology, 130, 156
Graber, D.A., 63
Grand Canyon Expedition (GCX), 112,
   152, 156–61
Grand Canyon National Park, xvi, 8, 31,
   37, 44, 115, 268; augmented reality field
   trip to, *160,* 152, 156–61
Grand Teton National Park A (GTNP), 50,
   76–77, 79, 110, *188;* Pura Vida
   program of, 168, 189–92
Great Smoky Mountains National Park,
   xvi, 37, 38, 50; Monarch Butterfly
   Tagging in, 93; rainfall data collection
   in, 97–107, 98
Greater Yellowstone Ecosystem, 80
Grenny, J., 25
Gross, M.P., 223
Group 4 Project, 76–77
Gruenewald, D.A., 140
Guam, xix
Guiden, Andrea, 112, 137–48
Gutwell, J.P., 230, 233, 234

habitats, xiv, 55, 75, 79, 172, 183
Haden, C.A., 230
Haefele, Michelle, 268
Hall, Ansel, 7

hallowed ground, xiv–xv, 2, 29–34
hands-on learning experiences, 79, 99, 196–98, 229, 233, 249
Harding, Vincent, 213, 214
Harlem Growth Youtn, 198
Harpers Ferry, 30; Center for Interpretive Media, 9; National Park, 54
Harvard University, Kennedy School of Government, xvii
Healthy Parks Healthy People (HPHP) initiative, 111–12, 127–33
heritage, engagement with, 24
Hertzog, George B., 9
Hess, J., 64
high school students, xvi, 67, 76–77, 80, 142, 168, 272. *See also* Mather High School; field trips for, xv, xvi 2, 30–34, 160, 152, 156–61, 272; parks as outdoor classrooms for, xiii–viii, 187, 190
historic preservation, 8, 32, 36, 169, 194, 196, 271. *See also specific historic sites*
historical content retention, 137, 144, 147
Hoh Rainforest, 58
Hollowell Park, 123
Houseal, Ana, xv, 50, 78, 83–93, 217, 221–39
Hurricane Frances, 100
Hurricane Ivan, 100
hypothesising, 76–78, 99, 107, 115, 121, 152

identity, 118, 225. 248, 266, 268
Illinois, xvii
immersive experiences, xiv, 15, 112, 153–56, 158, 161
impacts, xv, 11, 37, 49, 50, 73, 78, 112; of citizen science projects, 85, 93; of climate change, 54–67, 167, 172; of collaborative programs, 105–7; community, 73–76; geological, 152–53; on health, 128; of extreme weather, 100, 103; of place-based learning, xiv, xv, 30–34, 74, 77, 79–80, 115, 137–41, 144–46; of augmented and virtual reality field trips, xiv, 112, 153–56, 158, 161
inclusion, 17, 18, 190, 232, 237
Indiana Dunes National Lakeshore, xvi
Indianapolis, Children's Museum of, 239
informal learning, xx, 80, 227, 233
inquiry, 74, 80, 137, 230, 233–34, 236, 248; into the past, 24–25; scientific, 85–87, 233, 251

Inside Lakes, 39
Institute of Museum and Library Services (IMLS), 246, 247
Integrated Precipitation and Hydrology Experiment (IPHEx), 101
interactive learning, 2, 6, 142, 144, *150*, 155, 159–61; curriculum for, 169, 203; games in, 233, 234
interdisciplinary approach, 55, 74, 80, 112, 141, 172, 249; crosscutting concepts in, 86, *87,* 89, *91*
International Baccalaureate (IB) world schools, 76
International Dark Sky Parks, 123
internet, 30, 227, 261. *See also* social media
internships, 140, 197–98. 260
Interpretation and Education Renaissance Action Plan, 15–16
interpretation, xv, 7–18, 38, 137, 183, 213, 272; environmental, 10; evolution of, 18; for families, 237–38; of historic sites, 10–11, 112, 140–48, 203–12; as management tool, 11; partnerships in, 14, 17, 172–73, 212–13, 237; strategies for, 15, 172, 249
Interpreting Our Heritage (Tilden), 9
Interpretive Development Program (IDP), 13, 15
Interpretive Skills Vision Paper (Advisory Board Education Committee), 17–18
interviews, 49, 53, 54, 60, 62, 202, 211; with families, 223–24, 227; students, 141–43, 145–47, 210
intrinsic motivation, 117, 120
invasive species, 50, 54, 56, 75–76, 80, 152, 172
Iroquois Indians, 30
Islam, 210

Jackson Hole, 73, 80
Jacobs, Joshua, 196–97
Japanese-Americans, internment of, xv
Jarvis, Jon, xv, 33, 196, 271–73
Jefferson, Thomas, 2, 30
Journey Through Hallowed Ground (JTHG), xv, 30–34, 112, 272
Junior Nature School, 8
Junior Rangers, 137, 143, 195, 201, 235

Katmai National Park, xvi
Kawuneeche Valley, 123
Kelly Warm Springs, 75–76, 80
Kenai Fjords National Park, 53–54, 56, 58, 59

Kern, Amanda, 49, 73–80
King, Martin Luthr, Jr., 143, 146, 147, 214
Klondike Gold Rush National Historic Park, 220
Knutson, K., 233
Krasnow, Kevin, 49, 73–80
K-12 schools, xv, 50, 80, 83, 86, 177, 214; children's experiences outside, 266; curriculum for, 12, 33, 203; partnerships with, 206, 212

landscape managment, 169, 196
Larsen, David, 13
Latinx, 141, 168, 189–91, 210, 227
Laurel Gap, 96
leadership, 73–80, 207–9, 213, 250; heritage, 1, 23–26; National Park Service, 7, 11–16, 18, 31, 49, 55, 174, 183, 195; in partnerships, 168, 185–86, 190, 198
learner-centered education, 245, 272
Learning Historic Places with Diverse Populations (View and Azevedo), 202–14
Learning Science in Informal Environment: People, Places and Pursuits (National Research Council), 266
Leary, Patrick, 49, 73–80
Lent, Lotte, 17
Leopold, A., 25
lesson plans, xvi, 13, 34. See also curricula
LGBTQ community, 271
libraries, xiv, 8, 18, 33, 246, 247, 254
Lieberman, Gerald A., iv
lifelong learning, xx, 16, 86, 18, 57, 174, 248
Lincoln Memorial, 22
literature reviews, 54, 202, 223, 249
Little Rock Central High School National Historic Site, 143–44, 147
Loomis, John, 218, 266–67

Machlis, Gary, 272
Madison, James, 30
Making Sense of History: Understanding Landscape Change in Alaska, 59
Mammoth Hot Springs, 90
Manassas National Battlefield, 31
Manzanar National Park, xv
March on Washington (1963), 207
Marin headlands, 260
Marlowe, Tim, 218, 266–67
Marshall, George C., 32

Maryland, 30, 112
Massachustts, University of, xvii
Mather High School, 169, 194, 195–99
Mayo, Corky, 12, 13
McCain, John, 54–56
McClennen, Nate, 49, 73–80
McDonough, Colleen, 112, 151–61
McKay, Joe, 46
McMillan, Donna K., xvii, 111, 115–23
meadows, 15, 42, 43, 55
Mensesini, Mario, 10
Meterological Instrumentation, 105
Mexico, 155
middle schools, xiv, 31, 169, 190, 210–11; field trips for students of, xv, 30–34, 112, 137–48, 272
Milky Way, 122–23
Miller, Douglas K., xvi, 50, 97–107
Mills, Enos, 6–7
Minnesota, 116
Mission 66, 8
Missouri, 77
modern life, 86, 121–23. See also twenty-first century
Monarch Butterfly Tagging, 93
Monocacy National Battlefield, 32
Monroe, James, 30
Montana, 46; University of, 39
Montezuma National Park, xvii
Monticello National Park, 2, 30m 32
Monument Valley, 268
motivation, xv, 61, 64, 121, 127, 150–51, 161, 266; academic, 245–47, 252; in family visits, 225–27, 235–36; intrinsic, 117, 120
Mott, William Penn, Jr., 11
Mountain Raingers, 96, 97–107
Mount Guyot, 102
Mount Rainier National Park, 54, 57
Mount Sterling Ridge, 102, 103
Muir, John, xvi, 6, 37, 152
Muir Woods National Monument, 260
museums, xiv, 5, 6, 8, 11, 18, 138. See also specific museums; free-choice learning in, 61, 217; family learning in, 220, 221–39; partnerships with, 14; self-directed exhibits in, 141
Muslims, 211

Nagle, Lara, 112, 127–33
National Aeronautic and Space Administration (NASA), xvi, 98, 99, 101, 105
National Capital Parks East, 54, 60

National Civil Rights Museum, 141, 143, 144, 146
National Council for Interpretation and Education Volunteers, 17
National Council for Social Studies, 13
National Council of Negro Women (NCNW), 205, 207
National Environmental Education Development (NEED), 10
National Environmental Satellite, Data, and Information Service (NESDIS), 101, 105
National Geographic, 32
National Geographic Society, xv, xvi, 16, 93, 199
National Heritage Areas, 30
National Historic Preservation Act (1966), 32
National Mall, 211
National Oceanic and Atmospheric Administration (NOAA), xvi, 98, 99
National Park Foundation (NPF), 12, 17, 67, 168, 173, 176–79, 246,; Learning Alliance, 250
National Parks Conservation Association, 16, 17
National Park Service (NPS), xiv–xvi, 5–19, 31, 38, 80, 119, 152, 167, 217. *See also specific parks;* Advisory Board Education Committee, xv, xix, xx, 6–7, 14, 17, 272; Centennial Celebration, 33, 167, 171–73, 245; in citizen science partnerships, 93; "Civil War to Civil Rights" sites of, 137–48; climate change concerns in, 49–50, 53–67; Comprehensive Surveys of visitors of, 174–76; Conservation Association, 6, 16; *Every Kid in a Park* (EKiP) program, 176–78, 222, 235, 271; family learning enhancement by, 221–39; *Find Your Park/Encuentra Tu Parque* campaign, 176–77, 179; free-choice learning valuation by, 265–69; Healthy Parks Healthy People program, 111–12, 127–33; Historic Architecture, Conservation, Engineering Center for Northeast, 196; Interpretation and Education Renaissance Action Plan, 15–16; Interpretive Development Program (IDP), 13, 15; Interpretive Operations, 11; Junior Ranger program, 137, 143, 195, 201, 235; legislation establishing, xix, 152, 195, 271; Mather Training Center, 9, 13, 17, 18; National Education Council, 15; National Leadership Council, 14; of New York Harbor, 196; Office of Environmental Interpretation, 10; uutcomes for environmental education by, 245–54; Parknet, 13; Parks as Classrooms, 12; Passports, 209; rangers at. *see* rangers; Second Century Commission, 16, 195; Strategic Plan of, 271; Teaching with Historic Places, 13; value of education and learning programs of, 259–63, 261
National Park Service Centennial Act (2016), xix, 18–19, 272
National Park Service Organic Act (1916), xix, 152, 195, 271
National Parks for the 21st Century: The Vail Agenda, 11–12
National Research Council (NRC), 266
National Science Educational Standards, 12
National Science Foundation Advancing Informal STEM Learning progam, 246
National Science Teachers Association, 12
National Trust for Historic Preservation, 32
Native Americans, 38. *See also* American Indians
NatureBridge, xv
Nature Conservancy NYC LEAF Program, 198
Nature of Science (NOS), 77, 78, 80, 87
neuroscience, xiv
Newton, Susan, 167–68, 171–80
New York City, National Parks Service high school in, 168–69, 194, 195–99, 272
New York Harbor, xvi; National Parks of, 196
New York State Education Department, 197, 198
Next Generation Science Standards (NGSS), 157, 249
Nez Perce National Park, iv
95% Solution, The (Falk and Dierking), 266
nonprofit organizations, xiv, 12, 14, 78, 80, 168, 184–87
North American Association for Environmental Education (NAAEE), 246, 250
North Carolina, 100–103; University of, xvi, 50, 97, 101, 102, 105
North Cascades Institute, 66
North Cascades National Park, 37, 54, 57, 66

Obama, Barach, 18, 53, 214, 272
Obama, Michelle, xvii

Obed Wild and Scenic River, *170, 178*
observation, 6, *48,* 75, 78–79, 154, 156,
    202, 224; in citizen science projects, 85,
    90, *91* ; weather, 98–101, 105–7
Ocean Beach, 260
Of the Student, By the Student, For the
    Student, 31
Ohio, 168, 183
Ohio River, xv
Old Faithful, 1, 5
Olympic National Park, 54, 57, 58
on-site programming, 260, 261
Oregon State University, 18
Orr, David, 64
outcomes, 15–16, 25, 93, 173, 245–54;
    academic, 73–74, 77–78, 107; crosscut-
    ting, 218, 224, 246, 249–50, 253–54;
    health, 127–28; educational, xv, 12–13,
    32, 61, 73–74, 80, 155, 195, 197–98,
    217–18, 245–54; in partnerships, 202–4
Outlands Historic House & Gardens, 32

Pacific Islanders, 271
Parks as Resources for Knowledge in
    Science (PARKS), 12
Parks, Rosa, 214
Pearl Harbor National Memorial, iv–xv
Peck, Greg, 49, 73–80
Pennsylvania, 30, 223–24
Percoco, James A., 2, 29–34, 140
personalized learning, 74
Philadelphia-area museums, 223–24
Pigeon River Basin, 100, 102
Piscataway Indians, 30
Pisgah National Forest, 102
place attachment, 58, 64, 66–67
place-based education, xiv, xx, 2, 49,
    73–79, 195–99, 214, 231; case studies
    of, 75–79, 141–42; about climage
    change, xvi, 55–56, 63–67; curriculum
    for, 50, 73; engagement in, 54, 73–74,
    142, 180; free-choice learning in,
    265–69; at historic sites, 33–34, 140,
    147–48; learner-centered, 245;
    partnerships in, 167–69, 171–80
    190–92, 259; philosophy of, 74–75;
    technology in, 16; youth development
    outcomes of, 248
Place-Based Education Evaluation
    Collaborative, 140
Powell, John Wesley, 152
Powell, Robert B., 217, 245–54
preservation, xix, 17, 64, 197. *See also*
    historic preservation

President's Park. *See* White House National
    Park
Presidio, 260
Prince William Forest, 54
programming, 1, 53, 106, 116, 171, 173;
    educational, 7–8, 10–14, 18, 49,
    246–54, 259–60, 261, 267; health, 129,
    131–32; in partnerships, 176–78, 180,
    184
project-based learning, 74, 80, 196
Public Participation in Science Research, 84
Puerto Rico, xix
Pullman National Historic Park, xvii
Pura Vida program, 168, 189–92

race, xx, 23, 112, 130, 146, 174–75, 202
Racial and Ethnic Diversity of National
    Park System Visitors and Non-Visitors
    (Taylor, Grandjean, & Gramann), 174
rainforests, 58, 184
Ranger-Teacher-Ranger program, 212
Rawls, John, 23
Reilly, Patti, 12
Renewing Our Education Mission (National
    Park Service), 14–15
residential learning programs, xv, 14, 84,
    168, 183
Rethinking the National Parks for the 21st
    Century (National Park System
    Advisory Board), 14
Revisiting Leopold (NPS), 171–73
Reynolds, General John, 29
Ridenour, James, 11
Rising Wolf, 39–46
Rock Creek Park, 4
Rocky Mountain National Park, *xxii,* 6, 54,
    56–58, 63, 66, 111, *114,* 115;
    Environmental Psychology at, 116–17;
    120–23; types of learning at, *119*
Roosevelt, Eleanor, 208

St. Olaf College, xvii
Salazar, Ken, 196
Samoa, xix
Sánchez, Jésus, 168
Schoodic Point SeaWatch, 93
schools. *See* elementary schools, high-
    schools, K-12 schools, middle schools
Scientific and Engineering Practices (SEPs),
    86–90, *91,* 92
Scripps Institution of Oceanography, 101
sea level rise, xvi, 56, 65
self determination and efficacy, 77, 78, 120,
    248, 252

Selma to Montgomery National Historic
    Trail, 38, 143–44
Senate, U.S., Energy and Natural Resources
    Subcommittee, 54
service learning, 31, 67, 80
*Servicewide Interdisciplinary Strategic Plan
    for Interpretation, Education and
    Volunteers* (NPS), 249
Shanley, Deborah, 168
signage, healthy behavior influence by, 112,
    127–33
Six Americas, 65–66
smartphones, 30, 151
Smithsonian Institution, 247–48
Snake Den Ridge Trail, 102
Snake River, 75
Snapchat, 151
social fabrics, xviii, 64
social media, 13, 85, 105, 132, 151, 177,
    179, 213
sociocultural context, 226, 226, 228,
    236–37
Southwest Alaska Network (SWAN), 59
Spotsylvania, Battle of, 30
Sprague Lake, 55
Stanton, Robert, 14
Statue of Liberty National Monument, 9
Stegner, Wallace, iv, 273
STEM, xx, 154, 158, 246, 249, 254, 246,
    249
Stephen T. Mather Building Arts and
    Craftsmanship High School. *See* Mather
    High School
Stern, Marc J., 217, 245–54
stewardship, 12, 17–18, 171–72, 179, 211,
    222; environmental, 198, 246, 249, 253
Stones River Cemetery, 147
Storer College, 9
Storksdieck, Martin, 219, 225, 265–69
strategies, 15, 18, 49, 205, 272; education-
    al, xix–xx, 33, 86, 217–18, 249,
    266–69; engagement, 171–76, 178–80;
    health improvement, 112, 127–28, 133
Student-Scientist-Partnerships, 84
Student-Teacher-Scientist Partnerships, 84,
    90
Student Transfer Reform Act (California,
    2010), 157
Susquehannock Indians, 30
Swannanoa River, 100
Sweet Grass Hills, 39

tablets, 30, 151, 155
Taff, B. Derrick, 112, 127–33

*Take the Journey: Teaching American
    History Through Place-Based Learning*
    (Percoco), 34
Taking Science to School (National
    Research Council), 88
Tblisi Declaration (1977), 247
Teacher-Ranger-Teacher program, xvi,
    212
technology, xviii, 15, 23, 24, 121, 155, 174.
    See also STEM; family-friendly, 174,
    238; observational, 101, 156, 160; in
    place-based learning, 16
Tennessee, 100, 102, 178
terrorism, xviii
Teton Science Schools (TSS), 49–50, 168,
    272; Journey School, 76, 77; Kelly
    Campus, 77; Pura Vida program, 168,
    189–92
Thompson, Jessica, xv, 49, 53–67
three-dimensitonal learning, 83–93
Tilden, Freeman, 8–9, 140, 142–43, 213
Tomales Bay, 260
Tredegar Iron Works, 147
Tule National Park, xv
Tuskegee Airmen National Historic Site,
    143, 144, 147
Tuskegee Institute, 147
twenty-first century, xiii, 17, 198, 254;
    skills in, 247–49, 252
Two Bears, 41–42, 45

Udall, Mark, 54, 55
undergraduates. *See* college students
Underground Railroad, xvi, 30
underrepresented groups, 191, 223, 227
underwater research, xvi
UNEP, 247
United Kingdom, 233
United Nations Educational, Scientific
    and Cultural Organization (UNESCO),
    247
United Nations Environment Program
    (UNEP), 247
Upper Geyser Basin, 1, 5
US Geological Survey, 152
USS Arizona Memorial, iv
Utah State University, 77

View, Jenice L., 112, 137–48, 169, 201–14
Virgin Islands, xix
Virginia, 2, 30, 32, 33, 226
virtual experiences, 17, 122, 154, 156, 160,
    203; field trips, xiv, 33, 112, 152, 155,
    161

Vision Paper: 21st Century National Park Service Interpretive Skills (NPS-ABEC), 249

War of 1812, 30
Washburn, Julia, xv, 1, 5–18
Washington, D.C., xix, 54, 56, 60, 129, 169; teacher-ranger professional development project in, 201–14
water, xvii, 38; clean, xiv
Watson, Michael, 11
Wave Hill Forest Project, 197
Weaver, Bruce, 38
webcams, xvi, 13
West Virginia, 30, 115
White House National Park, xvii, 200, 205, 207–11
Wilderness, Battle of the, 30
wildlife, 38, 50, 58, 61, 115, 118, 172; viewing, 55, 72, 116
Wildlife Expeditions, 72, 78
Wimpey, Jeremy, 112, 127–33

Winks, Robin, 5
Wirth, Conrad, 8
World of Life (California Science Center), 268
World War II, 8
Wright, Katie, 168
Writing-on-Stone Provincial Park, 39
Wyatt, Cate Magennis, 31, 32
Wyoming, 78, 79, 189; University of, 77

Yale Project on Climate Change Communication (YPCCC), 65
Yandala, Deb, 168
Yard, Robert Sterling, 6
Yellowstone National Park, iv, xix, 1, 5, 37, 77, 79, 115, 225–26. *See also* Expedition Yellowstone!
Yevya, Kathy, 12
Yosemite National Park, iv–xvi, 9, 37, 250; Free Nature Guide Service, 7
Yosemite Valley, 152

Zion National Park, 37, 44

Founded in 1893,
UNIVERSITY OF CALIFORNIA PRESS
publishes bold, progressive books and journals
on topics in the arts, humanities, social sciences,
and natural sciences—with a focus on social
justice issues—that inspire thought and action
among readers worldwide.

The UC PRESS FOUNDATION
raises funds to uphold the press's vital role
as an independent, nonprofit publisher, and
receives philanthropic support from a wide
range of individuals and institutions—and from
committed readers like you. To learn more, visit
ucpress.edu/supportus.

Milton Keynes UK
Ingram Content Group UK Ltd.
UKHW011042020624
443508UK00005B/204